# Global Perspectives on Science and Spirituality

# Global Perspectives on Science and Spirituality

EDITED BY PRANAB DAS

TEMPLETON PRESS

Templeton Press
300 Conshohocken State Road, Suite 550
West Conshohocken, PA 19428
www.templetonpress.org



Designed and typeset by Kachergis Book Design

Library of Congress Cataloging-in-Publication Data
Global perspectives on science and spirituality / edited by Pranab Das.
p. cm.
Includes index.
ISBN-13: 978-1-59947-339-0 (alk. paper)
ISBN-10: 1-59947-339-9 (alk. paper)
1. Religion and science. I. Das, Pranab K. (Pranab Kumar)
BL240.3.G58 2009
201'.65—dc22
2009010159

Printed in the United States of America
09 10 11 12 13 14 10 9 8 7 6 5 4 3 2 1

# Contents

# Preface

**Pranab Das,** Elon University, United States

The Global Perspectives on Science and Spirituality program was launched in late 2003 with the purpose of identifying and supporting top thinkers outside the usual spectrum of science and religion research. Our goal from the outset was to participate with the excellent intellectual communities of Eastern Europe and Asia to bring fresh and invigorating input to the scholarly dialogue in the United States and Western Europe. The essays contained in this volume were contributed by team members from twelve of the best award-winning projects selected by that program, scholars drawn from a pool of more than 150 applicants in two dozen countries.

These essays will stimulate those familiar with the science and religion literature and engage general readers with an interest in the flux of ideas across disciplines, traditions, and regions. The contributions are richly textured and intellectually appealing. They offer entrée to new fields and starting points, both in the texts and in their references, from which to delve further into many fascinating questions. This volume is suitable for the interested lay reader and specialists looking to broaden their horizons. It is an accessible stand-alone text, but would also serve well as a source of additional readings in courses focused on science and spirituality in its many forms.

I am extremely grateful for the support given to this project by the John Templeton Foundation and its staff, especially

Dr. Paul Wason, without whose thoughtful and consistent involvement the GPSS would surely never have gotten off the ground. My warm and sincere thanks go to my colleagues and partners at our European center of operations, UIP's Tom Mackenzie and Jean Staune. Tom's extensive groundwork and contact-building were essential to developing the outstanding pool from which we chose our awardees. Thanks also to our judging panel that worked diligently to select the very best of a highly qualified group of applicants. I acknowledge the support of Elon University, which made time and resources available to me for this project. And most importantly, I wish to thank the love of my life, Dr. Kate Fowkes, for her generosity, patience, and endurance in the face of the many long voyages and international hours that the GPSS demanded.

# Introduction

**Pranab Das,** Elon University, United States

The book in your hands is an invitation. It is not a collection of lengthy monographs or comprehensive scholarly treatises. Nor is it a compendium of worldwide perspectives on a single theme or topic in science and religion. Rather, it is a collection of scintillating overtures—introductory thoughts that open doors onto new terrains. Spanning ten countries and many spiritual and scientific backgrounds, the thinkers whose work is represented here are actively engaged in exceptional research, the exploration of science and spirituality from distinctly non-Western perspectives.

Each essay displays approaches distinctly different from the mainstream analyses of Western scholarship but nonetheless well rooted in the contemporary American-European literature. They embrace the paradoxes of Eastern thought. They deploy terms of ancient and polyvalent significance that promise years of unpacking. And they offer the science and religion dialogue community exciting new opportunities and challenging new paradigms.

The contributors to this volume are extraordinary scholars. They have each achieved a pinnacle in their national academies. Professors at flagship universities like India's JNU and the Czech Republic's Charles University, and at research centers like the Nanzan Institute for Religion and Culture, the Russian Academy of Sciences' Institute of Philosophy, and the

Indian National Institute of Advanced Study, they are also path breakers willing to bridge the chasms between academic disciplines. These are interdisciplinarians of the highest level who have combined excellence in philosophy, theology, and the sciences with unique and thrilling results. And they have committed their exceptional minds to work that will inform the broader intellectual community while knowing full well the danger that their immediate disciplinary colleagues may lack the interest or inquiring spirit to appreciate the breadth of their work.

These authors remain aware of the need to avoid compounding disciplinary suspicion with obscure or jargon-laden prose. They have each constructed a brief, accessible essay that will, we hope, entice the reader into a larger dialogue with the scholarly context within which they move. Every essay is readable and compact, and each makes a case for re-visioning the boundaries around analytic modes; each one appeals to our innate excitement at uncovering the hidden connectedness between our disparate thought structures.

Our Asian contributors offer breathtakingly fresh ideas to the science and religion discourse. Their deep familiarity with Western scholarship is combined with access to whole canons of literary, philosophical, spiritual, and religious material that have been largely under-researched in the West. Some, like Makarand Paranjape, link to these resources along paths that are familiar to Western scholars. Paranjape contextualizes his essay within the complex literature on colonialism. India, having been an example both of colonial impact and postcolonial emergence, seems at first to offer an easy case study. But Paranjape makes it clear that this radically heterogeneous culture presents a similarly diverse set of responses to modernity and, in fact, can be thought of as a truly *amodern* context. Hence, the unique Indian worldview, with its tremendous wealth of foundational thought, presents a deeply attractive locale for novel and productive interactions between modern science and spirituality.

Ryusei Takeda unveils the subtleties of Buddhist thought through a close read of a Western intellectual icon, Alfred North Whitehead. Whitehead, whose "process" thought is the foundation of many of today's most intriguing approaches to the theology of science and technology, directly confronted Buddhism. Though his critique was cogent, Takeda argues that he overlooked its essential dynamism, a dynamism sometimes masked by overemphasis on the stillness of the meditative mind. In fact, it is the creative coming-into-being that identifies the "discrimination of nondiscrimination," that most elusive and important stage of intellectual involvement with the world that has been aspired to by centuries of adepts and sages. Process, though construed in very different ways, is actually at the core of both an enlightened Christian theology and Buddhist thought. And it is through an understanding of the unfolding universe that the developmental world of science and technology comes into focus from a spiritual perspective.

Heup Young Kim frames the meeting of East and West in a more specific but no less trenchant way. Writing in the aftermath of a South Korean cloning researcher's notoriously exaggerated claims, Kim takes aim at some of our essential assumptions regarding human rights. He offers an overview of how Western ethics fails to span the space of "being in relation" that typifies Confucian virtue, a failure based in the occidental emphasis on self-actualization. Uniquely positioned as both a Christian theologian and a highly trained Confucian scholar, Kim marks out a space where the fundamental disconnect between our Western philosophies of science and applied ethics run aground on ancient and fully mature Eastern spiritual practices and traditions.

Sangeetha Menon leaps directly to the cutting edge of science, the tantalizing question of human consciousness. She cogently presents the key reductionist and emergentist perspectives from Western academia but slashes through the Gordian knot that they collectively present with sharp application of Sanskrit terms deeply imbricated with the centuries-old practices and philosophies of the subcontinent. Her introduction to the fascinating potential of "pure I-ness" and the noncontradiction of Indian conceptual systems is indeed an invitation to dive headlong into what offers to be a revolution in our understanding of spiritual human being.

For the group headed by Paul Swanson at Nanzan Institute in Japan, one word alone is enough to engender a hugely expansive idea system. In the term *kokoro* they find a portal into the core of science and spirituality. Spanning concepts like "mind," "heart," and "soul," *kokoro* resonates in every Japanese analysis of how things and beings interact and how vital existences be and become. In brief introductions we meet experts in robotics and climate analysis, primatology and neurology, all of whom perceive this one word as a prism through which to glimpse the real potential for interconnected thinking about life, science, and technology. If they succeed in fully unpacking this delightfully involuted concept, we all stand to gain a brand-new perspective on many important issues.

Among the essays from Eastern Europe are ruminations on the nature of mathematical thinking and its deep impact on the structure of science. Alexei Chernyakov and Botond Gaál, thinkers whose careers and memories are marked by the closed ideation of communist rule, wrestle with the hermeneutic implications of openness in mathematics, the seemingly most rigid of analytical structures. They find that modern mathematics is rooted in change, that its own capacity for reinvention is the underlayment of contemporary science and the key to its vast successes. And they ask if that same ongoing process of self-reinvention might not be the key to progress in modern Christian theology.

Ladislav Kvasz takes this idea further by arguing that transcendence is not only the meeting ground between science and religion but the essential common thread that runs through the most enlightening ideas of each. He reaches into the guts of conventional theories of science and religion, scientific revolutions and theology to emerge with a coherent and challenging new concordance that builds bridges across multiple levels of thought.

Ilya Kasavin and Anton Markoš likewise ground their work in rigorous philosophy from the continental tradition. But they arc out toward grander visions of human potential and the very essence of life as such. The science-religion dialogue may be a route toward a more humane future where the technologized context of human living is recontextualized and revalued. Or we may be able to apply a semiotic analysis to the biome, and so situate ourselves in an actively communicating living world that urgently seeks self- and other-knowledge as it undergoes a constant process of dynamic development.

Markoš' biosemiotics marks traditional categories such as "evolution" and "creation" as obsolete simplifications. Grzegorz Bugajak and Jacek Tomczyk arrive at a similar conclusion from within the Catholic tradition in Poland, a country now beginning to face the evolution-creation feud that has washed across Western landscapes for so long. They point out the essential inconsistencies in every attempt to define human specialness from a purely reductionist perspective and note that only by emphasizing the primary reality of *process* can we reasonably encompass the depth of human being with due reverence.

The overarching plan of this volume was to feature short overviews, lures to the inquiring reader to delve further into the work of very accomplished scholars. The contributions are generally not conclusive. They offer many more questions than answers, openings that will spark lively conversation and deep reflection. It is with eager anticipation of new contacts, cross-pollination, and deeper exploration that we offer this collection, *Global Perspectives on Science and Spirituality*.

Global Perspectives on Science and Spirituality

Editor's Introduction to

# The Puzzle of Consciousness and Experiential Primacy: Agency in Cognitive Sciences and Spiritual Experiences

**Sangeetha Menon,** National Institute of
Advanced Studies, India

Dr. Menon unpacks the subtle differences between consciousness itself and the system that is responsible for consciousness in a unique and compelling argument spanning mainstream Western research and the Indian spiritual discourse. The so-called "binding problem" arises from a dualistic juxtaposition of highly distributed neural function and the coherent "selfness" of human experience. Menon focuses her attention in this chapter on the challenging analytic problem of translating the first-person experience of consciousness into the necessarily third-person constructs of scientific epistemology. Calling this problem "harder" (in homage to Chalmers' well-known typology of the "hard problem"), Menon acknowledges an epistemological breakdown that masks the ontological essence of consciousness itself.

Rather than take refuge in the traditional categories of reductionist analysis, emergentist methodology, or a retreat to a purely subjective testimonial sense of self, Menon prepares here a combination of approaches that grows out of the rich Eastern wisdom traditions. In Indian philosophical and spiritual practice, the epistemological rupture is dealt with by crafting "distinct but interrelated ontologies." By applying metaphorical and narrative strategies, a relational understanding is achieved even in the absence of a structured set of classifications and relations between neural or cognitive objects. Menon makes a crucial point in debunking a common misconception related to this epistemological strategy. The Indic emphasis on transcendental experiences, she points out, does not imply a division between the ordinary experience of life and spiritual experience (or the intimate self-knowledge of meditative states). Rather, the self, one's identity, responds in an integrated way to a nondual experience of "I-consciousness."

Menon brings these perspectives to bear on the foundational concept of agency. While different from the primary self, Menon argues that "agency and experience are connected in an integral way." She proposes to employ three elements of Indian spiritual study as contexts in which to explore the "integrality" of consciousness and agency. These elements, characterized by the three terms *bhakti, jnana,* and *natya,* are seen to meet in a transcognitive superstructure that spans "spiritual agency" and the uniquely Indic concept of "pure I-ness."

As a preliminary investigation, this chapter acknowledges the deep challenges in reanalyzing these central questions in light of Indian scholarly and spiritual dialectics. The very translation of her three terms as love, knowledge, and aesthetic experience, for example, creates significant losses by stripping important subtlety from them and locating them in a Western epistemological framework. Furthermore, Menon is aware of the Vedantic prescription, mirroring essential semiotic theory, that it is the combination of the act of knowing and the known that is the knower. Addressing the classic Godelian paradox, Menon proposes that the analytic tools of the Indic tradition offer a nonlinear approach to understanding consciousness and an acknowledgment of the essential value of "just being."

P.D.

1

# The Puzzle of Consciousness and Experiential Primacy

## Agency in Cognitive Sciences and Spiritual Experiences

**Sangeetha Menon,** National Institute of
Advanced Studies, India

### Introduction

The major epistemological worry faced by the empiricist, the philosopher, and the psychologist centers on "experience." This worry is finding a theoretical explanation for the mutual influence of neural events and subjective experiences, which is the defining characteristic of consciousness.

Interestingly, any attempt to understand "experience," such as simple physical pain or complex psychological pain, will have to cross epistemological barriers of hierarchies and causal relationships. The classical description of consciousness as "unitary" has evolved to accommodate the questions emerging in interdisciplinary dialogues, to present the term "self," which was once considered purely metaphysical but today is very much available for scientific discussion. The epistemological transition, however implicit, is from a third-person perspective to first-person and second-person perspectives.

A distinctive trend in "consciousness" discussions started

with the theory of "easy problems and hard problem" by David Chalmers (Chalmers 1995), which for the first time in the Western world made a semantic distinction between "being conscious" and "what is responsible for consciousness" and presented the challenging "binding problem."[1] Both experimental and cognitive sciences acknowledged the strong presence of an "explanatory gap" (Jackendoff 1987). Though their approaches still remained/remain reductionistic (or at least dualistic), the complexity of "consciousness" and its unique nature were largely accepted. This acceptance inspired theories favoring complex cognitive and social functions, neural and subneural structures, system-environment interaction, etc., in order to fill the "explanatory gap" and place "consciousness" in its seat.

The views that are currently discussed and debated no longer fall into a strict division of reductionistic and nonreductionistic approaches. This could be because of the growing recognition of a distinct characteristic of "consciousness," namely that it is not strictly linear. Another reason is the need to bridge the first-person and third-person worlds. One prominent view is that there is a distinction between subjective conscious experience and the biological mechanisms responsible for these and that they are mutually nonreducible. This view is based on the position that first-person data cannot be fully understood in terms of third-person data (Chalmers 2000). Biological explanations have also factored a hierarchy of functions in order to explain consciousness. One such view holds that consciousness is a highly complex motor response occupying "the uppermost echelon of a hierarchy having the primitive reflex at its base" and that which "arises from the systems' interactions with the environment" (Cotterill 2001).

Approaches explaining consciousness as epiphenomenal, but not in the classical sense of emerging from a physical composite, also take into account the fact that the primary problem is more than a theoretical divide between the empirical and the subjective aspects of consciousness. Some of these approaches hold that consciousness "is formed in the dynamic interrelation of self and other, and therefore is inherently intersubjective" (Thompson 2001) or that it is a system of interactions between the animal and its environment and that it is not located in the brain (Varela et al. 1991). Explanations that address the psychological and social dimensions of consciousness hold that consciousness is "some pattern of activity in neurons" (Churchland 1997) or that it is best understood in terms of varying degrees of "intentionality" (Dennett 1991), and in terms of "memes," which are the units of cultural evolution (Dawkins 1976; Blackmore 1999).

Yet another school of thought that calls for finding neural correlates for the subjective components of consciousness focuses on the scientific exploration of meditation techniques. This school acknowledges the contribution of Eastern philosophy and wisdom traditions in developing an understanding of the use of meditation techniques for transcendental and extraordinary states of consciousness and experiences (Varela and Shear 1999).

It is interesting that many of these scholars of thought consider consciousness as a phenomenon to be *understood*—that it is within the same scope of investigation and dialogue as any other phenomenon. There is a degree of equal balance between two basic explanations/approaches for consciousness such as (a) neural/physical/social correlates and (b) extraordinary and meditation (transcendental) experiences mostly validated by neural or other third-person data.

## The Puzzle of Experience

The questions we ask about consciousness have their bases on different kinds of experience such as dreams, states of mind, memory, pain (physical and mental), etc. But the analysis of these questions is based on segregated information about behavior or brain events and processes. Therefore, the answers to these questions are given in terms of neural correlates and neural information processing and models thereof. This method strips away the essential aspect of "being conscious" or "consciousness," which is the "person." Questions asked as a result of first-person experience are answered based on third-person information. Essentially, there is a gap between the problematic of conscious experience and the attempts to address it, which I call the "harder problem" (Menon 2001). The standards and criteria that we follow for objective understanding are most often the criteria for third-person information. This method helps us to build technologies and to understand abnormalities transcending individual existences. The first-person qualitative methods give us opportunities to be sensitive to individual nature, psychology, expression, and uniqueness.

If both methods are important, how can the "harder problem" be addressed? I do not have a ready answer for this question. But we could attempt a method that is neither mutually converting (information to experience and experience to information), reductionistic, nor solipsistic. We should avoid the presumption that the larger picture of consciousness emerges out of solely third-person or first-person methods. The "harder problem" is not a question, I think, to be answered completely, or a complete theory about consciousness. Rather, it is the ontological essence of "consciousness" that should always be addressed by whichever method we adopt. This will help us to *see* beyond the third-person information and the first-person experience.

The availability of "consciousness" for our most intimate experiences and our simultaneous inability to understand it *completely* in terms of third-person information makes us think that "consciousness" is a complex phenomenon and that its complexity needs to be addressed. We understand "complexity" as an intrinsic characteristic of the "other," the object of investigation. This notion of ours about "complexity" should be reexamined. Simple methods may reduce the many features of a phenomenon to a manageable number, but such methods are not complete third-person representations of first-person phenomena. Complexity could be the characteristic feature needed for the design to provide full third-person representation.[2]

It should therefore be noted that the standard scientific criterion of replicability cannot be applied since the third-person representation cannot be a replica of the complete first-person phenomenon but only a *representation* of it from a particular framework that follows certain epistemological and empirical/theoretical parameters.

### Experiential Intimacy

According to the Chalmersian theory of "easy" and "hard" problems, first-person data cannot be subjected to the standard method of reductive explanation. This theory also questions the basic fact of consciousness: why is the performance of neural functions accompanied by subjective experience?

The "why" question here is pertinent to the bases on which we find our primary, secondary, and tertiary questions and methods for understanding "consciousness." The "why" question ("why neural functions are accompanied by subjective experience") assumes:

1. consciousness as a separate "something" borne or unusual/nonnatural,
2. (neural) functions as basically having only mechanistic meanings, and
3. subjective experience as not the intrinsic nature of consciousness.

These assumptions, indirectly upheld by the camp of antireductionism, stem from the basic conflict between "experience" and "cognition." The normative criteria for establishing "truth" start with the objective reduction of whatever is posited. Subjective experience fails to pass the normative tests as agreed-upon, valid data. So the why-question arises from the conflict between epistemological necessity and experiential primacy. Both seem to be unavoidable and coexistent in human discourse.

It is difficult, if not impossible, to resolve a conflict if both sides are equally important. But the recognition of this unavoidable conflict in our theories and models will help us to widen the scope of investigation and prevent dehumanization of our goals. After all, through both third-person information and first-person experience what we ultimately seek are personal growth and health, coexistence and sharing, and a continuous exploration into the unknown and the unpredictable.

Among the key elements in our agenda is the primary division between the meaning and scope of "awareness of something" and "awareness by itself." What exactly is "self awareness"? It is awareness of something. It is either the awareness of:

1. the world outside, such as other states of mind, objects, etc., or
2. the world inside, such as "my emotions," "my perceptions," "my body," "my identity," etc.

The "world inside" cannot be understood without the intervention of self-reflection and self-participation. What is "awareness itself"? Awareness itself can be seen as:

1. uniting discrete thoughts, and the two worlds (inside and outside),
2. as meta-awareness of the two (inside and outside world) awarenesses,
3. as pure I-ness.

All these are necessary to make consciousness available for *experiential intimacy.*

Unless a clear distinction is made between these different categories of existence, we will end up searching for the needle in the same haystack for centuries without realizing that the problem is not only the subtleness of the subject of our inquiry but our own inability to design a comprehensive search. The design of the comprehensive search is important because the way we search alters the very presence of our quarry in that invincible heap.

**Indian Routes for Dialogues**

"Consciousness" has become the umbrella term for many issues crossing and connecting disciplines. The route and the possible result of this dialogue is to connect and join various streams of thought, whether empirical or intuitive, experimental or theoretical, in order to map and locate consciousness. On a scale of meta-analysis, this is a linear and horizontal approach because our dialogues start from third-person working definitions of "consciousness" (however different they are). Despite the variety and differences in the themes and ideas for human discourse, "consciousness" studies have been an attempt to harmonize and integrate otherwise divergent human thinking.

The Eastern wisdom traditions, beginning from the Vedic system of thought, perceive entities (physical/metaphysical) that connect the outside world of objects and the inside world of experiences. There are several verses in the *Brahmanas* that speak to the quest for the source of knowledge and experience. These are characteristic of Indian philosophy from its origins to its classical schools and saints where even the most realist schools have focused on uniting all the units of the self rather than beginning from the outside and working in.

Isolated epistemological analysis, in Indian thought, is subservient to experiential paradigms. Indian schools of thought, in general, have one common thread—that is to relate to a larger, deeper, and holistic concept/entity called "self." Whether it is for affirmation or denial, Indian thought engages in rich analytic thinking to form a philosophy about "self." Both analysis (structured and a "leading-to-next" kind of hierarchical thinking) and experience are used as epistemological tools in an integral manner to form distinct but interrelated ontologies. Metaphors and imageries become epistemological tools for creating transcendence in thinking and thereby experiencing. The aim is not to arrive at structured and classified/listed knowledge of an*other* object/phenomenon but to understand relations with an abiding entity whether it be the self/no-self/matter.

Another interesting feature of Indian philosophical thinking is the importance given to the way of living or lifestyle subscribed to by the schools, no matter how realistic or idealistic their metaphysical position is. The understanding of a particular school of thought will not be fulfilled by "understanding" its epistemology or even worldview but by following a lifestyle that is prescribed. Experience is the core of understanding. This requires the student's mind to follow certain rules and to take on the discipline of forming integral and interrelated connections rather than individual and isolated relationships. This is a major difference with the dialogues in the West on "consciousness."

Indian philosophy elevates the therapeutic value of analysis and the self-oriented integration of understanding over its cognitive value. The initial and final reference points for the inquiring mind is the "person" and his/her experience and the situation he/she is in. The route taken is from the situation of the person (as "given") to the reorientation and reorganization of his/her response based on transpersonal experiences. The challenge is to strive to make from what is given (Menon 2003). Hence the style of discourse adopted in their presentations is more metaphorical and nonlinear than hierarchical and localized.

The classical approach to spiritual experiences is to disengage from "ordinary" experiences and engage "transcendental" experiences. This is sometimes misunderstood as implying that there is a division between (and a need to travel from) the "ordinary" to the "transcendental" experience. This misconception stems from failing to understand the primary thesis that spiritual experience is not another kind of experience in another world and relating to another set of objects, forsaking and condemning the "ordinary" experienced world as something of a hierarchically lower order. Spiritual experience, according to the Advaita, is a reorientation and, thereby, a reconstruction of any experience from the Self's point of view. The ontological thesis of Sankaracarya upholds I-consciousness as "something-which-is-already-there." Spiritual experience is a reconstructing of any experience from this ontology. "It is there across, above, below, full, existence, knowledge, bliss, non-dual, infinite, eternal and one" (Sankaracarya 1987).

The difference between an "ordinary" experience and a "spiritual" experience is that in the former case the experience is given meaning from the point of view of self and in the latter case it is understood from the point of view of Self. In both experiences there is an identity that relates to and generates meanings. In the first case the identity is caused and defined by the situation. In the latter case the identity defines the situation by responding to it from an integral point of view.

This thesis has been incorrectly criticized as an *ad hoc* rationalization for the overarching philosophy's lack of soteriology. Similarly, consciousness as described by Sankaracarya is often mistaken as *niskriya* (inactive) in its literal sense. The notion of *maya* too has invited many misconceptions, the most common suggesting a passive homogeneity to pure consciousness. The main argument behind such misconceptions can be traced back to the labeling of Advaita as monistic.

Current discussions on “consciousness” mostly focus on one of the two problems—how simple physiological functions coordinate and work together as one single system, and how and why a subjective orientation ensues. Would a focus on the ontology of Self and human experience reveal a different picture of consciousness? In the first case the attempt is to build into “consciousness” and in the second case the attempt is to build from “Self.” The categories of thought needed for the two cases are different. One is for the allocation of new knowledge within a system, and the other is for transformation of knowledge asystemically. Experience is the common concern of and mystery in each discussion (though it is not the beginning point for the first approach). But can we give up the experiential primacy of consciousness totally or give it secondary importance? This would seem not to be so easy and, more importantly, such a move results in less meaningful results.

## Agency and Experience

Recent discussion on “experience” embraces the fact that to understand consciousness, even with the help of case studies of abnormal conditions, is to factor in the nonquantifiable components of human possibilities such as self-effort and a positive outlook. Ongoing discussions that address the “unity” of consciousness, especially in the field of cognitive science, suggest two phenomena as significant to understanding the binding nature of consciousness. These phenomena are “intention” and “agency.” This leap forward in cognitive science took place half a decade ago with the shift of focus from “a biological organ for consciousness” to “a *situated* brain” (Menon 2005).

However, the scientific study of agency remains in its infancy and cognitive science has started off with a minimalist conceptual definition of the term. Agency is defined as the sense of authorship of one’s actions (Kircher and Leube 2003) and is also known as the “Who System” with underlying cognitive processes (Georgieff and Jeannerod 1998). Due to the difficulty of adopting a first-person methodological component, the essential expressions of “raw feelings,” called “primary self-experiences" in cognitive science, are not included in the conceptualization of agency.[3] This could be the reason that much of the discussion on agency in the context of consciousness and experience is limited to sensory-motor activities and impairments. Another significant debate of recent years has focused on "altruism and meme machine" (Blackmore 1999), a debate that has undergone many revisions in its framework and philosophical positions over time. I have argued that deeper exploration of the complexity of agency might also contribute to appreciating not just biological selfishness but also the basics of "spiritual altruism" (Menon 2007).

How different or related are self and agency? Agency and the primary sense of self are phenomena that seem to be inextricably intertwined with an individual first-person perspective of consciousness. Experience requires both an authorship and a primary self. Further, experience is made meaningful by the continued "feeling" of self. A sense of enduring and pervasive identity is the starting point of all experiences. Agency and experience are thus connected in an integral way.

The experience of agency is most significant not only in the formation of a first-person perspective but also in the development of social cognition, intersubjective values such as empathy, and coexistence. The possibility of transformation and transcendence in human experiences continuously reminds us of yet another significant dimension of human consciousness—spiritual experience (Bodhananda and Menon 2003). How we should address spiritual experience from the point of view of understanding consciousness in the context of agency is a significant question. It is necessary to explore how the "integrality" provided by agency and the effacement of selfhood in spiritual experience could be correlated.

The three contexts of spiritual experience from the Indian discourse—love (*bhakti*), knowledge (*jnana*), and aesthetic experience (*natya*)—give a picture of the complexity of agency and its intricate relation with consciousness, experience, and self. This discussion is expected to initiate a review of agency from the point of view of diverse disciplines and spiritual experiences.

## Conclusion

Agency has a transcognitive function. In the above three instances of spiritual experiences, this idea is presented as pure I-ness. Pure I-ness can be related to a new idea about authorship (namely "spiritual agency") that will help us to look at agency not only as a factor that helps to engage but also to detach. This is the meeting point between consciousness and agency, as conceived in current scientific discussions, and the three spiritual experiences described from the Indian discourse in the context of love (*bhakti*), knowledge (*jnana*), and aesthetic experience (*rasa*).

To understand agency within a positive framework of the possibility of human experiences, a distinct methodology and research program will have to be developed that will enable us to accommodate both the third-person content of experience and the first-person agency of consciousness. Such a method might first help to analyze and then resolve various phases of agency on an acausal and apodictic level. The narrative modes of conscious experience (Petranker 2003) might help us to balance between the fringes of pure subjectivity and pure objectivity. Studies in spiritual neuroscience that indicate the nonlocal, spread-out neural correlates of mystical experience (Beauregarda and Paquettea 2006) and the significance of reviewing various neurocognitive views (Morin 2006) will help us explore not just experiential primacy (Menon 2005) but the *spiritual intimacy* implicit in consciousness.

*Agency* will be an important tool in the coming years to explore the nature of consciousness and also to understand how much subjectivity can be extracted from consciousness and how much objectivity can be attributed to it. A major obstacle to purely objective research on consciousness is the fact that at any given moment it is impossible to observe the observation of observation since all the three cannot be simultaneous. With a third-person approach we can capture only a frame-by-frame rendering of the event. The content of consciousness, meanwhile, undergoes a series of changes so that each frame is redefined and reconstituted in the act of "capture." Such a static kind of tool, and the metaphors that accompany it, may not do full justice to a phenomenon like consciousness. This could be the reason that the "stream" and "flow" metaphor has been widely used by authors from William James (1950) to Mihalyi Csikszentmihaly (1991).

Since observation *per se* requires another observation to become an object of inquiry at any given time, there will always be a subjective component to experiment, a component impervious to the act of observation. At the same time the knower is not just intricately but indivisibly connected with observation so that we can know only the content of observation. Even if we manage to isolate the knower from the known content, the knower will only have a contextual relation—namely with the known content.

This is the reason why, in the Vedantic treatises, consciousness by itself is beyond the triad of the knower, the act of knowing, and the known. The triad of objective discreteness is sustained by consciousness without "itself" being a part of the triad. The final isolation of knower will present not the cognitive knower but its "original" state (being) of pure I-ness. What is experienced in a spiritual experience is a glimpse of this reality.

The dynamics of the levels of agency and their relation to consciousness can be better appreciated through the study of three contexts of spiritual experience from the Indian discourse: *natya* with the focus on body as the primary tool, *bhakti* with the focus on mind as the primary tool, and *atmajnana* with the focus on reflection as the primary tool for transcendence.

A nearly complete theory of consciousness can be developed with the help of a nonlinear mode of participative inquiry. A final understanding of consciousness might best lie in the transformation of self and the adopting of the complex yet spiritual nature of agency. Both factors bring us back to the issue of *experience* as the means as well as result of understanding consciousness. As Aristotle said, "Understanding is the understanding of understanding" (*Metaphysics* 11.10741). Abhinavagupta, the tenth-century Indian aesthete, lauded "the gift of heart" (*Tantraloka* 3:200),[4] and Michael Polanyi, in more recent times, emphasized the significance of "passions" in discovering "truth" (*The Personal Knowledge* 1958),[5] perhaps the harder problem of consciousness extends from a qualia-centric neural agency to a being-centric experiential and spiritual intimacy. Finally, understanding will have to give way to another mode of knowing and that could be *just being*, which seems to be the greatest challenge for all of us.


### Acknowledgments

I thank my colleagues at the National Institute of Advanced Studies and friends for their critical reviews and discussions. I thank Sambodh Centre for Living Values for the spiritual space that helped me understand the meaning of spiritual agency.


### Notes

1. "Binding experiences" is how physical, discrete, quantitative neural processes and functions give rise to experiences that are nonphysical, subjective, unitary, and qualitative.
2. Perhaps what we distinguish as "simple" and "complex" are not the intrinsic characteristics of the object of investigation, but the categories of thinking and understanding we have formed according to the third-person information supplied to us by the tools we have designed. Hence, the question, "Should design and tool be complex?" becomes important.
3. These studies could be supplemented with the intervention of the study of spiritual experiences that might give another dimension to the cognitive notion of agency. The question whether spiritual experience is exclusive and disconnected from our day-to-day living or is it something that is foundational to how we engage with day-to-day living is a significant issue for discussion, which this study hopes to engage in.
4. *tatha hi madhure gite sparse va candanadike*
   *madhyathi avagame yasau hrdaye spandanamanata*
   *anandasaktih saivokta yatah sahrdayoh janah*
   *Tantraloka* 3:200

   When the ears are filled with the sound of sweet song or the nostrils with the scent of sandalwood, etc., the state of indifference (nonparticipation) disappears and the heart is invaded by a state of vibration; such a state is precisely the so-called power of beatitude, thanks to which human is "gifted with heart."
5. According to Michael Polanyi we should start from the fact that "we can know more than we can tell." This prelogical phase of knowing was termed "tacit knowledge" by him.

## Bibliography

Beauregarda, Mario, and Vincent Paquettea. "Neural Correlates of a Mystical Experience in Carmelite Nuns." *Neuroscience Letters* 405, no. 3 (2006): 186–90.

Blackmore, Susan. *The Meme Machine.* Oxford: Oxford University Press, 1999.

Bodhananda, Swami and Sangeetha Menon. *Dialogues: Philosopher Meets Seer.* New Delhi: BlueJay Books, 2003.

Chalmers, David. "The Puzzle of Conscious Experience," *Scientific American*, December 1995, 62–68.

———. *The Conscious Mind: In Search of a Fundamental Theory*. Oxford: Oxford University Press, 2000.

Churchland, Paul S., and T. J. Sejnowski. *The Computational Brain*. Cambridge: MIT Press, 1997.

Cotterill, Rodney M. J. "Evolution, Cognition and Consciousness." *Journal of Consciousness Studies* 8, no. 2 (2001): 4.

Dawkins, Richard. *The Selfish Gene*. Oxford: Oxford University Press, 1976.

Dennett, Daniel. *Consciousness Explained*. London: Penguin Books, 1991.

Dwivedi, R. C., and Navjivan Rastogi. *Tantraloka of Abhinavagupta*. Freemont, CA: Asian Humanities Press, 1985.

Georgieff, N., and M. Jeannerod. "Beyond Consciousness of External Reality: A "Who" System for Consciousness of Action and Self-consciousness." *Consciousness and Cognition* 7 (1998): 465–77.

Jackendoff, Ray S. *Consciousness and the Computational Mind*. Cambridge, MA: MIT Press, 1987.

Kircher, Tilo T. J., and Dirk T. Leube. "Self-consciousness, Self-agency and Schizophrenia," *Consciousness and Cognition* 12, no. 4 (2003): 656–69.

Menon, Sangeetha. "Basics of Spiritual Altruism." *The Journal of Transpersonal Psychology* 39, no. 2 (2007): 137–52.

———. "Binding Experiences for a First Person Approach: Looking at Indian Ways of Thinking (*Darsana*) and Acting (*Natya*) in the Context of Current Discussions on 'Consciousness.'" In *On Mind and Consciousness*, edited by Chhanda Chakraborti, Manas K Mandal, and Rimi B Chatterjee, 90–117. Kharagpur: Indian Institute of Advanced Study, Shimla and Department of Humanities and Social Sciences Indian Institute of Technology, 2003.

———. "Cognition, Consciousness, and Experience: Towards a New Epistemology." In *History and Philosophy of Science* PHISPC-CONSSAVY/vol. 2, part VI. Delhi: Indian Council of Philosophical Research, 2005.

———. "Towards a Sankarite Approach to Consciousness Studies." *Journal of Indian Council of Philosophical Research* 18, no. 1 (2001): 95–111.

Morin, Alain. "Levels of Consciousness and Self-awareness: A Comparison and Integration of Various Neurocognitive Views." *Consciousness and Cognition* 15 (2006): 358–71.

Petranker, Jack. "Inhabiting consious experience: Engaged objectivity in the first-

person study of consciousness." *Journal of Consciousness Studies* 10, no. 12 (2003): 3–23.
Polanyi, Michael. *Personal Knowledge: Towards a Post Critical Philosophy.* London: Routledge, 1958.
Ross, W. D. *Metaphysics*, http://classics.mit.edu/Aristotle/metaphysics.html.
Sankaracarya, Adi. *Atmabodha*. Translated by Swami Chinmayananda. Bombay: Central Chinmaya Mission Trust, 1987.
Thompson, Evan. "Empathy and Consciousness." *Journal of Consciousness Studies* 8, nos. 5–7 (2001): 1.
Varela, Francisco, Evan Thompson, and E. Rosch. *The Embodied Mind: Cognitive Science and Human Experience*. Cambridge: MIT Press, 1991.
Varela, Francisco, and Jonathan Shear, ed. *The View from Within: First Person Approaches to Consciousness*. Exeter, U.K.: Imprint Academic, 1999.

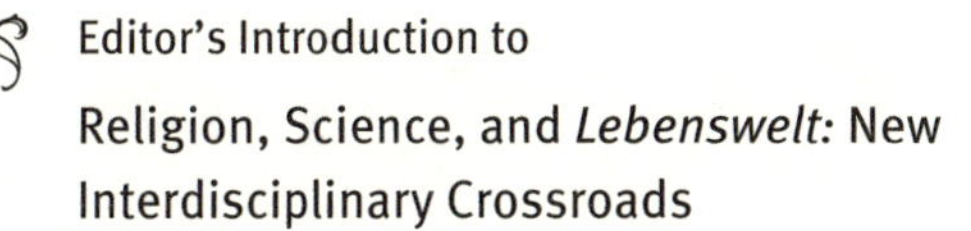

Editor's Introduction to

# Religion, Science, and *Lebenswelt:* New Interdisciplinary Crossroads

**Ilya Kasavin,** Institute of Philosophy, Russian Academy of Sciences, Russia

Professor Kasavin confronts the structural distinctions between scientific practice and the metaphysical explorations of philosophy and theology. As a senior scholar at the pinnacle of the Russian intellectual establishment, Kasavin has wrestled with sudden and dramatic paradigm shifts, and he draws from that experience to suggest that a coherent, guided interaction between the epistemologies of science and religion can enhance intellectual and spiritual freedom and "dethrone" conventional cultural forms.

Recent trends in academic dialogues have radically shifted the structure of the sciences and the humanities to the extent that C. P. Snow's famous two cultures are now irrelevant. In their place, a new compatibility is emerging in the philosophies of scholarship that privileges interdisciplinarity and acknowledges the insufficiency of disciplinary epistemologies. Kasavin argues that "postmodernist methodologies" that flexibly call on varied intellectual resources depending on context also span important changes in contemporary Christian thought. He traces the evolution of human cognition through religion and situates "artful" questioning of the chaotic universe as the philosophical leap forward of our time, a move that embraces a new area of study—science and spirituality.

Similarly, Kasavin sees the history of science as an overlay on the progress of human thought characterized by a movement from less complex to more enlightening constructs. And he finds a close parallel between the juxtaposition of the sacred and the profane and the analytic modes of theory and experiment. Citing the consequences of Russian sociopolitical transformation, he foresees a novel post-scientific thought system in which science is not completely deprivileged but perspectives are expanded.

The context for public living, the *Lebenswelt*, has been profoundly transformed by modern science but, most recently, the singular role of science and technology in guiding those changes is receding into the background and supporting a set of new modes of human interaction. Such technologies as the Internet provide the foundation upon which a new anomie is explored by the broader culture of "hypercommunication" and attacked by widespread *thanatos* (the Freudian term for "an aggressive urge to explore the limits of one's possibilities"). Kasavin points to two mutually contradictory tendencies in this technologized world. On the one hand, life's sense of meaning is jeopardized by increasingly technological and scientific overlays. But on the other, science itself if humanized and culturally located, thereby draining it of some of its authoritative sway over sociocultural dynamics.

Kasavin argues that it is in the interdisciplinary field of science and spirituality that a historically accurate recapitulation of the history of mind can be found. He argues that by embracing a flexible, evolutionary model of transdisciplinary analysis, it is possible to move toward "an essential universality." And he holds that the lens of science and spirituality offers "rational solutions to some significant intellectual controversies of our time."

P.D.

2

# Religion, Science, and *Lebenswelt*

## New Interdisciplinary Crossroads

**Ilya Kasavin,** Institute of Philosophy, Russian Academy of Sciences, Russia

The second half of the twentieth century saw such intellectual shifts and trends as postpositivism, postmodernism, posthumanism, etc. One of the essential features uniting all of these movements is a new understanding of science, on one hand, and of culture, on the other. The old division between "two cultures" introduced by C. P. Snow seems to have become irrelevant. Science loses its authority as the highest kind of objective knowledge and integrates nonscientific (political, commercial, religious, artistic) elements. And other cultural forms reveal their nonsubjective, that is impersonal, purely informational, mass-media character. All this leads to dramatic consequences for both fields whose identities diffuse and decrease to the extent that we begin to see pronouncements like "the death of Author," "the death of God," "the end of Culture," "the end of Science."

In this setting, a new area of research has been taking shape in the humanities[1] for the past thirty years. It lies on the boundary of two seemingly incompatible disciplines—the philosophy of science and the philosophy of religion. For various reasons, physicists, biologists, psychologists, sociologists, philosophers, theologians, and specialists in religious studies have become convinced that they should transcend their narrow disciplinary approaches. Scientists seek in religion a wider context for their research; theologians need scientific arguments to modernize the religious outlook; specialists in religious studies borrow sophisticated methodological approaches from philosophers of science; and philosophers develop a comparative analysis of cultures and speculate on the possibilities for a global synthesis of ideas. As soon as a new worldview is needed, it becomes clear that each discipline by itself is, in principle, insufficient. This need is embedded in the twentieth-century culture, the dynamism of which is unrivaled.

This development is marked, first, by the incessant generation and change of cultural models, and second, by new interdisciplinary approaches based on certain nontraditional theoretical presuppositions. These approaches include the social history of science, transpersonal psychology, interpretative anthropology, discourse-analysis, qualitative studies, along with others that relate to so-called "postmodernist" methodology. They generally share a pluralist view of culture as an aggregate of various intellectual resources that they combine freely for different purposes. The use and interpretation of resources depend on their current cultural and social contexts. For this reason, science and religion, knowledge and belief, logic and rhetoric, theories and metaphors, facts and fantasies can successfully interact and form unified conceptual systems. The Christian church has played an essential role in providing institutional support for this movement as it finds itself in the process of intellectual and organizational reformation. Today there is widespread discussion of whether a new discipline, "science and spirituality," should be introduced into lay education and into the structure of institutional science. Different religious institutions and charitable foundations offer a number of related scholarly and cultural programs to scientists and scholars interested in topics outside their academic areas with an aim of exploring this new field.

What are the conceptual and methodological presuppositions of such a discipline? It is not the first time we face this question. The same issue arose with cultural studies, political sciences, the science of science, the comparative study of religion, synergetics,[2] etc. In each of these cases, a particular type of research with an undefined disciplinary status brings together various subject areas, theoretical presuppositions, and methodological principles and procedures. The philosopher's task is to analyze some basic concepts and to reconsider their meaning and interaction. The notions in question are science and religion, knowledge and belief, research and revelation, proof and persuasion. Yet, as becomes clear at a closer view, these notions have constantly evolved and have been defined in concrete historical and sociocultural contexts.

### Rethinking Religion

We find in the Bible at least two concepts of faith. The first kind is the faith of Abraham, which is based on an equal contract with God, and the second kind is the faith of Job, which resulted from painful suffering, disappointment, and hope, that is, from self-knowledge. Abraham's faith comes from the world, where adequate knowledge and efficient action, including communication with God, are possible. By contrast, Job's faith follows from the breakup of the contracts that set both the rules for communication between tribes and the natural laws of events. Abraham's faith, which is grounded in sacrifice, cedes place to Job's faith, which comes accompanied by three "discoveries." The first is that the world loses its unshakeable order and allows for miracles according to God's will. The second is that Man loses the ability to understand and respond to God and to be in a dialogue with him. But, thirdly, as Carl Jung pointed out ("*Antwort zu Job*"), humiliated and suffering Man discovers in himself the ability for sophisticated reflection, which God does not have. He overcomes his suffering and becomes superior even to God. The faith in God as the source of an unchangeable natural and social order gives way to the faith in God's power and Man's freedom from the social contract. Man believes that he is selected and that faith makes him powerful. The mode of relationship based on the unshakeable contract is therefore questioned: "The traders are thrown out of the Temple," and the *contract* gives way to *faith as it is*.

Religion is a stage in the evolution of humanity's cognitive capacities, a stage that is necessary both rationally and historically and without which there would be neither philosophy nor science. One may distinguish within its periods, dominated either by the cult of propitiation of gods or by magic activism, and draw some inspiring conclusions from this difference. It is worth recalling the case of magic in the recent history of the philosophy of science. It was not accidental that at the end of the 1960s a discussion about the rationality of magic attracted much interest in the philosophy of science.[3] The crisis of the neopositivist tradition led to the rethinking of the concept "rationality" and to the introduction of nonscientific modes of thought into discussions within the philosophy of science. A very slow and gradual process of reconceptualizing religion, magic, and myth went along with the transition to the postpositivist philosophy of science. In particular, it became clear that the current life of magic extended far beyond its cultivation by the professed witch or sorcerer. Magic as a model of extreme experience is practiced by every creative agent who constructs an ontology of his own to transcend and to transform everyday reality. The analysis of historical forms of magic opened a door to the understanding of a more general problem—the problem of creativity. There is a creative action in the basis of every epic narrative: either a linguistic one, given in a form of literary writing, or a practical one, presented as a historical act. This act becomes at first a myth about gods and heroes, their mythical *arché*, then transforms itself into tradition and ritual, and, finally, turns into the subject matter of poetics that reproduces arché with the help of the magic of words.

The scientific revolution of the New Times begins when magic activism replaces propitiation, and "the conquest of the world" replaces the "concordance with being." The moral life no longer copies stable principles of social order but begins an endless spiritual growth. The harmonious law-abiding Cosmos gives way to a chaotic and diverse nature that "likes hiding." Man is no longer satisfied by the knowledge of its regularities, as set by the original contract, but wants to "question nature" in an independent and artful way. This particular change brings philosophy and epistemology into the center of attention. As a result of secularization, cognition and knowledge are transferred outside religion (for the most part) and into science and philosophy. Similarly, contemporary studies of religion no longer look like confessional theology; they have become an interdisciplinary study of a specific kind of knowledge, which is both practical and spiritual, connected both with activity and communication, and analyzed with the help of sociological, psychological, and epistemological methods and approaches.

### Rethinking Science

During the last five centuries, science also underwent a long evolution. To understand its structure and functions, one needs to know its genesis. In this sense, one can draw a distinction between three definitions of science. If we accept that Euclid and Archimedes created the sciences of mathematics and mechanics, Leibniz and Newton could only have continued what had already been started in antiquity. In this case, science is identical to a "scientific idea" or "scientific theory." By contrast, if we believe that science was born in the early modern period and that antiquity was no more than a propaedeutic, then by "science" we mean an intellectual movement that is based on a certain paradigm and requires certain systems of education and publication of results. In yet further contrast, one might assume that science, in the contemporary meaning of the term, appeared only in the mid-nineteenth century, because it was only at that time that the intellectual movement acquired an institutional base, funding pattern, specialized education, and independence from religion and politics.

A similar disagreement exists also about the origins of science: whether it emerges as a generalization of practice, a criticism of mythology,[4] a desacralization of magic,[5] a technical projection of a religious metaphysics, or as one of the ancient metaphysical speculations, in the form of abstract natural philosophy. Here, once again, science is taken in general as a whole, although it is a constellation of different phenomena such as abstract pure knowledge, naturalist research, technological systems, and social conceptions. It is obviously meaningless to look for their origins in the same place and in the same historical period. To do this would be to ignore the consensus view of science as a historically changing, diverse sociocultural phenomenon, a move that would be unacceptable even for those who analyze the inherent logic of science in the spirit of internalism.

Giving priority to cognition and examination, which is typical of science, is a relatively late historical development. Outside science—in myth, religion, and morality—the individual learns through assimilating collective representations or through spiritual growth. Historically, the pursuit of learning, which is present in magic,[6] then goes behind the stage in myth and religion and returns with philosophical and scientific knowledge. Having absorbed the results of prescientific and philosophical knowledge, the natural sciences developed one of the features typical of naturalist magic, namely *the search for and use of hidden powers of nature*. The opposition between the heavenly or divine world (regular, perfect, self-sufficient) and the earthly world (spontaneous, faulty, dependent) is based on other oppositions: order and chaos, cause and consequence, essence and appearance, law and fact, truth and error, the exact and the approximate. The heavens, with perfect motion, became a prototype for scientific theory, which emerged as a result of the philosophical and scientific rationalization of cognition process. The diverse and imperfect earth served as a prototype for empirical knowledge. The relationship between the *sacred* and *profane* is an ontological prototype for the epistemological relationship between the theoretical and empirical. The sacred cognitive attitude, dominant in prescientific naturalism, helped shape the norms and ideals of scientific knowledge such as truth, simplicity, exactitude, and objectivity, which would later become the standards of empirical research: repeatability, reproducibility, and observability. The laws of the heavens, superposed on earthly events, created the mathematical natural sciences, or the Scientific Revolution.

Conclusions reached about the relative and socially laden character of scientific truths, the negative consequences of scientific and technological progress, and a certain "saturation" with scientific issues have raised doubts about the special epistemological status of science and its intellectual power, doubts that threaten to deprive science of much of its funding. The American philosopher, Paul Feyerabend, treated science as a tradition with no more right to power than any other tradition.[7] Yet the interpretation of science as a tradition, together with the analysis of traditions in science, is a meaningful one because it sets science in various cultural contexts.

Contemporary research in the social history of science leads to the conclusion that science is as old as other achievements of humanity. A philosopher or historian of science should be sympathetic to the idea of expanding the concept of science and of legitimating forms of knowledge that are different from what we have been taught at school.[8] Yet today philosophers and scientists have already understood that human life, material as well as spiritual, is not limited to science and its applications. Moreover, this understanding in Russian public opinion is no longer in conflict with the dominant ideology. We can make one more step toward historical truth and examine the conventional and relative character of the terms "science" and "non-science"[9] in a way that by no means cancels the fact that science differs essentially from other cultural and intellectual institutions.

Were those who set up and solved the task of bringing ancient thought and culture back from oblivion, the lovers of the classics or, rather, scholars of the Cabbala and Hermeticism? There is no straightforward answer[10] because we cannot retroactively survey the intellectual communities that produced these results. We can, however, reflect upon who these classic authors of ancient philosophy were. They were not narrow people. Antiquity itself is far from being the kingdom of enlightened rationalism. That is why the dialogues like *Timaeus* and *Symposion* also inspired the occultists, and numerous pseudo-Aristotles were used in mystical metaphysics. Not the least because there are many interpretations, both apocryphal and genuine, of Plato and Aristotle, their ideas lived through the centuries and their authority was as great for Renaissance people as for the scholastics.

Thus, the birth of modern science, a phenomenon with which one usually associates the Renaissance and the early modern period, is an ambivalent process. The new cosmology (astronomy) owes not as much to the expansion of empirical observations and mathematical analysis of data as to the new worldview, which is a combination of rational and magical elements. To a similar extent, the next stage, classical mechanics, is connected with Platonism, heretical theology, alchemy, astrology, and cabalistic thought. The creator of the mechanist paradigm, Isaac Newton, lays in its foundation a tacit bomb, the theory of gravitation, which undermines the paradigm. A realization of the limitation of the Newtonian concept of the world, and a wave of interest in magical metaphysics and what today is called "paranormal phenomena," went hand in hand.[11] Up to now, science has not completely eliminated myth, magic, and religion: it *pushes them out*, to the sphere of alternative worldviews.[12] And as long as theory can serve instrumental and empirical practice, science can forget about the alternatives. A search for a broader outlook, including a search in the forgotten mystical and magical doctrines, coincides, as a rule, with periods of theoretical helplessness and disappointment, which are recurrent in the history of culture. Yet it is in these periods that chefs d'oeuvre are created, social utopias emerge, and scholars have great insights.

## Rethinking *Lebenswelt*

Accelerating scientific progress revealed the inevitable and immanent limits associated with scantiness of natural resources, political instability, social confrontations, and increasing instability of human mentality. Today modern natural science, which arose with the optimistic slogan of freeing the creative powers of man ("Knowledge is power"), is becoming one of the fields of venture as well as routine business. Thus it is losing its culture-constitutive function and generates numerous global problems. But the development of sociohumanitarian knowledge and counterscientific social movements contribute to the correction of this process. Thanks to them, alternative intellectual resources are being created and a formation of new paradigms in the field of natural and technical sciences is taking place. The integral elements of such paradigms become humanitarian expertise, social control, interdisciplinary cooperation, complex developing systems, and human-dimensional objects. Today the philosophy and theory of science project an image of science that is multidimensional, historic, socially and anthropologically laden, and in which pride of place is given to an analysis of the interactions among cognitive, psychological, cultural, and cosmological factors crucial for the development of science. It becomes apparent that many sciences do not completely eliminate certain life meanings as objects of analysis and as world outlook guidelines.

At the same time *Lebenswelt*, or the life world as a domain of everydayness, is transformed under the influence of civilizational reality to the point where it, just like modern science, is in need of a historic and nonclassical interpretation. For Husserl the life world consisted of initial, nonproblematic, and interrelated structures of consciousness characterized by integrity and stability. Such an understanding was preserved even in A. Schutz. But empirical research carried out by historians, sociologists, psychologists, and linguists substantially modifies this understanding. A historical changeability of the life world is being disclosed and that is why Husserl's phenomenology becomes limited with its depiction of only a classical type of the life world.

An attempt to interpret the life world of man of the twenty-first century as a field of a flexible and subjective reality, however, runs up against an unexpected obstacle. It turns out that the life world in question is, to a great extent, driven beyond the boundaries of the psyche. In the contemporary epoch, communication with technical devices comes to the foreground compared with communication with nature or among people (especially if we take into account the fact that the two latter types of communication are almost impossible today without technical devices). Most of what we do is switching from a computer to a telephone and from the telephone to the TV set. When we go out in the street, we can't do without a bus, a tram, metro, a car, a mobile phone, a transistor radio, a bicycle, roller skates, a baby carriage. . . . At the same time the role of these devices as tools moves to the background: a TV, a telephone, a car, a computer *as they are* become values independently from their utility in realization of the communication among people or between man and nature. On TV we watch programs about TV journalists; our telephone plays us electronic music that is downloaded into it; washing and fueling our car, buying parts for it and fixing it, discussing all this takes much more time than a car actually saves. And what can be said about personal computers, the potential of which is many years ahead of the needs of a common user?

The Internet is considered the peak of the informational revolution of our time. Many people view it as perhaps the most important development of the twentieth century, a development that broadens the capabilities of humans. It is true that the usual means of communication (telephone, post, telegraph) are slower and more expensive than Internet mailing or telephone systems. But let us try to go further and *question the quality of this communication and its subjects*. For someone who is deprived of a family and a lively conversation with friends or for someone who has physical disabilities, the Internet is an undoubted advantage that gives him or her access to the world. It also helps to *maintain* communication with friends, relatives, or loved ones when they are away. Everything else that the Internet has to offer to someone who is looking for communication remains within the limits offered by radio communication or by the "pen pal" practice that was popular among the youth in the 1960s who could not travel around the globe. For those who had a possibility to travel, the "pen pal" practice was a means of organizing cheap tourism. Very rarely did it lead to something bigger. The Internet today also gives an opportunity for anonymous acquaintances that do not last long, that impose no responsibilities, and that are very unlikely to lead to something bigger if they are to remain within the limits of the Internet.

Be it good or bad, the Internet is nothing more than a promotion of a new type of commerce. By making buying goods and services easier, the Internet rather enslaves a man than frees him. Therefore it is not contingent that trade (and especially the youth-oriented goods like video and audio devices, cars, tourist and sex services) takes up the central part of the Internet. The Internet's aim is to evoke a passion for obtaining the objects of pleasure and to engage him in the *use of the goods* of the modern civilization, thus making him a *professional user*.

According to sociologists[13] this technologization of communication is revealed in the superpenetrating hypercommunicativeness, or the media character, of the modern informational society. If everything conduces to the continuation of communication and the recursive inclusion of new messages in something already reported, if everything becomes communication, then there is no communication anymore, "it has died." In place of communication there remain streams of messages and monitors who are screening the viewers themselves. Thus the most important aspect of communication—reflection and understanding—disappears. And this means that the subject disappears as well from communication. Luhmann's concept of "Ego" (the addressee in communication) carries no substantial origin and serves only as a fiction or an operational scheme the function of which is to bring the chaos of experience into order. Luhmann's schemes form a nonreflexive mind of mass media—i.e., they form conditions of learning (and not of understanding and reflection). An event loses its novelty not in the sense that some time later there appears more fresh news. What is new cannot principally fit into mind; it is being missed by the mass media. The "terror of schematization" (J. Baudrillard) lies behind the hegemony of the production of sense.

A modern person in the circumstances of hypercommunication and instability paradoxically feels his or her loneliness in the crowd and the routine character of his or her being. And we find in those activities labeled by the jargon word "extreme" an alternative the purpose of which is to drag out a person from the swamp of his or her everydayness. So the domain of the life world is described in terms of *thanatos*. *Thanatos* (as introduced by Freud) is an aggressive urge to explore the limits of one's possibilities, the boundaries of what is socially permissible. Violation of a law, infidelity, sports, narcotic drugs, suicide attempts—all this is only a small part of the list of various actions that cut the peaceful flow of everydayness short for a certain time and sometimes even put an end to it forever (discontinuing at the same time the usual quality of life or life itself). Such phenomena acquire certain features of commonness no matter whether the people experiencing these phenomena want them to or not. The extent to which this happens depends on particular subcultures. But this is a different paranormal kind of everydayness. Modified forms of consciousness (due to extreme passions and affects or, on the contrary, due to low spirits, depressions, neuroses, or psychoses) correspond to this kind of everydayness.

Contemporary epistemological and interdisciplinary analysis of the life world shows its correlation with the stages in the development of scientific knowledge and institutionalized education. For a person representing a technogenic[14] culture, the skill of dealing with a computer and other technical devices, his affiliation with the streams of information and systems of communication, radically change his or her life world compared to the life world of his or her ancestors. And even though the structures of the life world function in many instances nonreflexively, their substantial difference is so great that it does not make for a problem-free understanding between people of different cultures and epochs. At the same time, certain differences within the life world (nation, language, and social strata) that used to hinder understanding become erased and thus certain conditions for a dialogue of cultures are created.

Therefore two self-contradictory tendencies realize themselves under the conditions of technogenic civilization. First of all, it is a technologization of the life world and an introduction of scientific concepts and practices into it. Both of these tendencies pose the danger of us losing a number of life meanings. At the same time a humanization and anthropomorphization of science turn out to be the reverse side of its vulgarization and a decrease of creative potential. But the world of science and humans' life world are not simply moved apart to different extremes. These are the extremes between which a constant meaning exchange takes place; they are poles, the existence of which ensures the dynamics of culture as well as the tension of philosophical discourse.

All this allows me to draw the following conclusions about the possibility and significance of science and spirituality as a special interdisciplinary area of research and teaching.

1. There is no sharp difference between science and spirituality in the historical genesis of all scientific disciplines. And a deeper understanding of the meaning of scientific terms requires their historical study.
2. The genetic view of any science is much more universal than a structural one, to the extent that evolution is not yet finished.

Science and spirituality represents, then, an essential universality.

3. Various cultural communities pay different attention to science and spirituality, so neither the former nor the latter in their isolation can play a solo role in intercultural understanding. The same is true in relation to a new world outlook, which has been gradually evolving in the last decades and which dominates the near future, marked by a confrontation between globalists and their rivals.
4. Philosophy of science and philosophy of religion develop through permanent exchange of methods and facts. The most impressive results of this interdisciplinary interaction are the ideas of method in theology (Bernard Lonergan) and of nonscientific knowledge (Kurt Hübner).
5. Science and spirituality as an area of research and teaching could be seen as exemplification of fairly general tendencies in the structure of scientific disciplines and in the culture in general. Today a number of decisive methodological and institutional shifts takes place in both spheres. Natural and exact sciences loose their *a priori* authority and interact with nonscientific forms of thought; social and human knowledge actively pretends to win scientific status; new, nonclassical sciences emerge; culture itself transforms into pop culture through technical innovations. In short, the very structure and proportions of *Lebenswelt*, the world of life, undergo radical transformations. All this creates an opportunity for science and spirituality and, even more, makes it a rational solution to some significant intellectual controversies of our time.

It happens sometimes that the philosophical image of nonscientific knowledge transforms itself from a challenge into a privileged research subject, which needs not a critical testing but blind faith. Then the tension between science and its setting, science and its alternatives, disappears, and the whole situation becomes the subject matter of religious studies or even a case for social criticism. The same happens when a scholar in the study of religion mixes the purposes of scientific analysis with the goals of ideological debates—it can be easily found out by a critical epistemologist. The similarity of interests for epistemology and religious studies is based upon an understanding of the significance of cultural multiplicity in the formation of our future world outlook. What are the genetic mechanisms of culture, the nature of social creativity directed toward the growth of spiritual freedom? How can cultural variety be described and typologized? How can the pretensions of particular cultural forms of intellectual and even political monopoly be critically dethroned? These are the tasks that can make the dialogue between philosophy of science and philosophy of religion especially fruitful.

## Notes

1. I will take a risk to underline that the possible disciplinary status if any of the "science and spirituality" is localized within the sphere of the social sciences and humanities. Even scientists as soon as they begin to contemplate this problem field leave the domain of natural sciences, go far beyond the pure "nature as it is" being immediately involved in the metadiscourse on knowledge, faith, belief, values, worldview, etc. In this sense, non- and postmodern natural sciences enter the stage of "humanization" and "humanitarization," and one of the clear indications of this is emergence and actualization of the "science and spirituality" controversy.
2. This "new science" elaborated by H. Haken was originally inspired by I. Prigogine (I. Prigogine and I. Stengers, *Dialog mit der Natur. Neue Wege naturwissenschaftlichen Deukens* [München: Piper Verlag, 1981]).
3. See C. Jarvie and Joseph Agassi, "The Problem of the Rationality of Magic." *The British Journal of Sociology* 18 (1967): 55–74. A number of eminent scholars and philosophers of science took part in the discussion—Steven Lukes, Tom Settle, J. H. M. Beattie, etc.
4. See K. Hübner, *Die Wahrheit des Mythos* (München: Beck Verlag, 1985).
5. See D. O'Keefe, *Stolen Lightning: The Social Theory of Magic* (Oxford: Martin Robertson, 1984).
6. See A. Elkin, *Aboriginal Men of High Degree* (Sydney: Inner Traditions International, 1977).
7. See P. Feyerabend, *Science in a Free Society* (London: New Left Books, 1980).
8. See Y. Elkana, ed., *Science and Cultures: Sociology of the Sciences*, vol. 5 (Dordrecht and Boston: D. Reidel Publishing Company, 1981).
9. For the results of the French scientist M. Gocelin on the scientific status of

astrology, see P. Grimm, ed., *Philosophy of Science and the Occult* (New York: SUNY Series in Philosophy, 1982).

10. See, for instance, Heinrich Cornelius Agrippa von Nettesheim, *Die magische Werke* (Wiesbaden: Fourier, 1982); Agrippa von Nettesheim, *Die Eitelkeit und Unsicherheit der Wissenschaften und die Verteidigungsschrift. Herausgegeben von Fritz Mauthner* (München: Dr. Martin Saendig oHG., 1969).
11. See the works of German Romanticism: Johann Wilhelm Ritter, *Fragmente aus dem Nachlasse eines junges Physikers: Ein Taschenbuch für Freude der Natur* (Heidelberg: Kiepenheuer Bücherei, 1810); G. H. Schubert, *Symbolik des Traums* (Bamberg: Kunz, 1814); J. Goerres, *Die christliche Mystik, 4 Baende,* Regensburg: Ratisbon (1836–42); J. Bernhart (Hrg.), *Mystik, Magie und Daemonologie, München* (1927); Helmut Werner (Hrg.), *Hinter der Welt ist Magie* (München: Diederichs, 1990); C. G. Carus, *Über Lebensmagnetismus* (Leipzig: Brockhaus, 1857); C. G. Carus, *Symbolik der menschlichen Gestalt* (Leipzig: Brockhaus, 1953).
12. For relevant discussions, see, for example, J. Taylor, *Science and the Supernatural* (London: Temple Smith, 1972); J. Alcock, *Parapsychology: Science or Magic?* (Oxford: Pergamon, 1981); L. Truzzi, "The Occult Revival as Popular Culture," *The Sociological Quarterly* 13 (1972); J. Ellul, *The New Demons* (New York: The Seabury Press, 1975).
13. See N. Luhmann, *Die Realität der Massmedien* (Opladen: Westdeutscher Verlag GmbH, 1996).
14. See V. Stepin, *Theoretical Knowledge*, vol. 326, Synthese Library (Dordrecht: Springer, 2005).

## Bibliography


Agrippa von Nettesheim, H. C. *Die Eitelkeit und Unsicherheit der Wissenschaften und die Verteidigungsschrift. Herausgegeben von Fritz Mauthner.* München: Dr. Martin Saendig oHG., 1969.

———. *Die magische Werke.* Wiesbaden: Fourier, 1982.

Alcock, J. *Parapsychology: Science or Magic?* Oxford: Pergamon, 1981.

Bernhart, J. (Hrg.). *Mystik, Magie und Daemonologie.* München, 1927.

Carus, C. G. *Symbolik der menschlichen Gestalt.* Leipzig: Brockhaus, 1953.

———. *Über Lebensmagnetismus.* Leipzig: Brockhaus, 1857.

Elkana, Y., ed. *Science and Cultures: Sociology of the Sciences.* Vol. 5. Dordrecht and Boston: D. Reidel Publishing Company, 1981.

Elkin, A. *Aboriginal Men of High Degree.* Sydney: Inner Traditions International, 1977.

Ellul, J. *The New Demons.* New York: The Seabury Press, 1975.

Feyerabend, P. *Science in a Free Society.* London: New Left Books, 1980.

Goerres, J. *Die christliche Mystik.* Regensburg: Ratisbon, 1836–42.

Grimm, P., ed. *Philosophy of Science and the Occult.* New York: SUNY Series in Philosophy, 1982.

Hübner, K. *Die Wahrheit des Mythos*. München: Beck Verlag, 1985.
Jarvie, C., and Joseph Agassi, "The Problem of the Rationality of Magic." *The British Journal of Sociology* 18 (1967): 55–74.
Luhmann, N. *Die Realität der Massmedien*. Opladen: Westdeutscher Verlag GmbH, 1996.
O'Keefe, D. *Stolen Lightning: The Social Theory of Magic*. Oxford: Martin Robertson, 1984.
Prigogine, I., and I. Stengers. *Dialog mit der Natur. Neue Wege naturwissenschaftlichen Deukens*. München: Piper Verlag, 1981.
Ritter, J. W. *Fragmente aus dem Nachlasse eines junges Physikers: Ein Taschenbuch für Freude der Natur.* Heidelberg: Kiepenheuer Bücherei, 1810.
Schubert, G. H. *Symbolik des Traums*. Bamberg: Kunz, 1814.
Stepin, V. *Theoretical Knowledge*. Vol. 326. Synthese Library. Dordrecht: Springer, 2005.
Taylor, J. *Science and the Supernatural*. London: Temple Smith, 1972.
Truzzi, L. "The Occult Revival as Popular Culture." *The Sociological Quarterly* 13 (1972).
Werner, H. (Hrg.). *Hinter der Welt ist Magie*. München: Diederichs, 1990.

## Editor's Introduction to
# Science and Spirituality in Modern India

**Makarand Paranjape,** Jawaharlal Nehru University, India

Western science and Indian spirituality intertwined in the emergence of the modern Indian state. Professor Makarand Paranjape explores the interaction of modern science with the long-standing inquiry traditions of indigenous practice and notes an important continuity in spiritual practice that embraces both knowledge systems. He makes the incisive observation that the conflict model that would emerge from a simple elision of the postcolonial critique and a history of Indian science incorrectly disregards the essential characteristic of Indian thought, its capacity to embrace "multiple, incommensurable systems."

In fact, the very notion of "traditional science" is problematically heterogeneous, a plurality that is present, though masked by convention, in British imperial science. Paranjape points out that the Orientalist construction of the history of science in India and the Hindu nationalist perspectives are equally weak analyses. On the one hand, Indian traditional science is seen as an ancient idea system long stagnant whose juxtaposition with modern science exposes its woeful inadequacy. On the other, the Vedas are imputed to have previsioned the achievements of today's research by thousands of years. It is Paranjape's contention that a graduated, diffusionist model can recognize power relationships between the hegemonist and recipient cultures while still privileging unique Indian capacities and idea structures. But to do so requires a clear vision of the defining characteristics of a distinctly Indian science. Paranjape claims that this is intimately bound up with the role of Indian spirituality.

Paranjape grounds his analysis on the notion of a unique Indian *episteme*, the Foucauldian "ground of knowledge or conditions that make the creation of knowledge possible." Summed up by the physicist Jagadish Chandra Bose, one element of such an episteme would be that underlying unity exists even in a multiplicity of phenomena. Another would be the eagerness, unique in the developing world, of India's embrace of modern science that exists as an almost illicit attraction for its otherness, what Paranjape cites as Kant's famous dictum "dare to know."

For Paranjape, India is "radically amodern." He sees coexistence between all four of Barbour's categories of relationship between science and spirituality in India, a reflection of its cultural proclivity to span oppositions and contrasts. Moreover, the modernist tendency to compartmentalize science and religion is undermined in an India whose spiritual leaders are by and large welcoming of scientific thought. Paranjape's India holds the potential to cradle a new dialogue between science and spirituality, a uniquely gifted culture with ancient and modern commitments toward scientific knowledge accelerating into an altogether new form of postmodernity.

P.D.

## 3

# Science and Spirituality in Modern India

**Makarand Paranjape,** Jawaharlal Nehru University, India

Modern science was introduced to India under the shadow of colonialism. This neither means that its progress in India was simply a matter of European discovery and imperial dissemination, nor that there was no "science" in India prior to the British conquest of India. However, what is important to observe is that between modern science and traditional science there was a marked disjunction, as there was between traditional knowledge and "English education." Because these gaps have still not been properly studied, let alone bridged, the history of modern science in India is inextricably linked with the history of colonialism as well. All the same, the trajectories of the two are neither coextensive nor coterminus. While colonialism rose, reached its peak, then declined and officially ended, modern science has enjoyed a steady and incremental rise since its inception. In fact, after independence its claims to an exalted social, political, and cultural status have risen dramatically, especially with the heavy investment and continuous monitoring of the Nehruvian state in its growth and development. In his introduction to *Science, Hegemony and Violence:*

*A Requiem for Modernity*, Ashis Nandy called science "a reason of state" and in *Another Reason: Science and the Imagination of Modern India*, Gyan Prakash labels the second part of this book "Science, Governmentality, and the State." Both authors regard science as very much a part of how the Indian state seeks to see or project itself, deriving legitimation and political advantage from it.

When we look instead at the development of modern Indian spirituality, we see that though it is also inextricably linked with the history of colonialism and, later, nationalism, its causal connections with the two are not all that direct or determinate. Under colonialism, traditional Indian spirituality encountered modern Western ideas, including modern science. Indian spirituality, though not necessarily as challenged as religious practices and dogmas were, had, nevertheless to reinvent itself, a process that still continues. One way that it coped was by identifying a distinct realm for its own functioning that had little to do either with colonialism or with modernity. Yet several Indian spiritual leaders, starting with Rammohun Roy, took an active interest in Western science. Sri Ramkrishna's disciple, Swami Vivekananda, best exemplifies not just the curiosity of Indian spirituality in respect to modern science, but its first well-articulated enthusiasm, even endorsement of science. From time to time, other spiritual masters such as Sri Aurobindo and Paramhansa Yogananda also continued this interest in and partial approval of modern science. At the same time, India's national struggle for independence, especially the majoritarian thrust of it, had a distinctly spiritual coloring. Not only was the Indian National Congress founded by Alan Octavian Hume, who was a theosophist, but many of its prominent leaders, especially Sri Aurobindo, and later, Mahatma Gandhi, had an overt interest in spirituality. Yet India as a secular state kept itself officially aloof from matters religious. Unlike science, which was a part of the state policy, spirituality, though a political force, was never authorized or recognized by the state. Therefore, when we come to the relationship between modern science and spirituality in India, we see not so much causal or direct connections, but subtle and covert connections.

One starting point for an inquiry into these connections is to ask what relationship modern science in India bears to what is now called traditional or indigenous science, and then ask the same question of Indian spirituality. We notice at once vital traces of continuity rather than disjunction between traditional and modern spirituality in India. In fact, the holistic, nondualistic orientation of traditional Indian knowledge systems does not allow us even to separate science and spirituality too clearly in premodern India. One might argue that this was also the case in pre-Enlightenment Europe. The fragmentation of knowledge and the ensuing proliferation of specializations is thus relatively new even to the West. In fact, this fragmentation is itself one of the constituents of modernity.

We thus have our first contrast between science and spirituality as systems of social construction and cultural authority in India. While modern science was seen as being alien and superior to traditional science, modern spirituality was seen as a natural outgrowth and flowering of traditional Indian spirituality. Yet what is perhaps even more interesting is how traditional Indian science and traditional Indian spirituality were closely, even intrinsically linked. For instance, while Ayurveda, as its name suggests, was described as the fifth Veda, thus not only linked to the Vedas but also shared its worldview and notions of wellness, modern, Western medicine was seen as wholly secular, if more effective in some cases. It is not that Indians in the nineteenth century dispensed with traditional medicine in favor of Western medicine. Both systems coexisted, but it was thought that when the disease worsened, the patients turned to Western (modern) medicine, which was then designated as "English" (*vilayati*) medicine. In fact, the belief that this turn signified that the patient's condition was critical also suggests that for nineteenth-century Indians, Western-style modernity was often the last, even fatal resort. The coexistence of multiple, incommensurable systems of medicine persist even today in India, though the dominance of Western (modern) medicine is far greater. But this plurality of knowledge systems is also characteristic of the metaphysical and epistemological multiplicity of modern India, a location in which we see the unresolved coexistence of contending systems of signification and meaning. Though not primary to their relationship, the issue of power cannot be ignored while exploring the science-spirituality dialogue.

As David Arnold observes, because traditional Indian sciences were heterogeneous and plural, it becomes difficult not only "to characterise Indian science as a whole but also to determine the precise nature of its interaction with the forms of science and technology emanating from the West by the late eighteenth and early nineteenth century" (2). One might extend this argument to contend that Western or modern science, though ostensibly unified by one universal "methodology," is actually heterogeneous and culturally conditioned as well. Thus, British imperial science, as Deepak Kumar and others have shown, had its own peculiarities and identifying traits that made it different not only from science practiced in Britain, but also from science in other European countries. For instance, colonial science was more descriptive and enumerative than theoretical or experimental. It was also more heavily invested in fields such as geology, plant biology, mining, and agricultural engineering, all of which had direct commercial value.

The other factor that complicates the story of the growth of modern science in India is the debate over whether it was a case of impact and reception as was largely thought of earlier, or of continuous interaction and exchange as the more recent and considered opinions declare. There is also the allied question of how to model the historiography of traditional Indian sciences. The earlier view in this regard, which followed the Orientalist construction, was one of high achievement, followed by decline and stagnation, ending with a new rejuvenation under the Western stimulus. Baber, for instance, identifies three narratives about precolonial science in India (15–17). The liberals and utilitarians among the colonial administrators held that there was nothing prior to the arrival of Western science; all of Indian history presented a tabula rasa as far as any useful knowledge was concerned. Then there were the better-inclined British Orientalists who believed that there was some good science in ancient India, but that it went into decline during the largely Muslim medieval ages. Hindu nationalists often mimicked the latter claims, but as with Swami Dayanada Saraswati, took them to more exaggerated, even ridiculous levels by claiming that modern scientific knowledge was implicit in the Vedas.

We now have a more informed view, seeing instead remarkable progress and interaction between Muslim and Hindu ideas and sciences during the medieval period of Muslim rule in India. The decentered nature of India's polity and the lack of sustained research to reconstruct its history of its science make it impossible to offer a coherent account, but it should be clear that any simplistic, reductive, or unidirectional model will be misleading. This applies as much to the advent of modernity in India, in which science played an important role, as it does to the development and growth of modern Indian spirituality.

George Basalla's influential and often-cited article, "The Spread of Western Science," offers a three-phase "diffusionist" model. In the first phase of colonial discovery, expansion and conquest, the non-European areas served as sources of scientific data. This may be termed the "contact phase." In consonance with the colonial interest in exploiting the natural resources of conquered territories, botany, zoology, and later, astronomy, geology, and geography were emphasized. While modern science was disseminated in various parts of the world during this phase, Basalla believes that only the advanced countries of Europe were able to assimilate this new information and knowledge, thereby transforming science in metropolitan centers in Britain, France, and Holland (1967: 611–22).

In the second phase, which he calls "colonial science," local scientific institutions start to appear, with the participation of local-born scientists. Basalla calls this a "dependent" science because it was controlled and directed by colonial authorities and imitated metropolitan models. This applied not only to countries like India directly colonized by the British but also to China, Japan, and the United States.

Extra-European societies in the third phase strove to establish national or independent scientific traditions. Political independence, but more significantly, institutions of national importance, awards, state funding, and infrastructure brought scientific research to critical thresholds in a number of countries, in fact, enabling them, in some cases, to overtake European science. Both America and Russia achieved this stage during the world wars, while Canada, Australia, and Japan followed. The rest of the world in Asia, Africa, and Latin America lagged far behind.

Basalla has been universally criticized for being simplistic, unidirectional, and reductive but, after all, he was only offering a preliminary schema in a short paper. In fact, the categories he proposed, including "colonial science," have proved to be extremely influential and persistent. As Arnold sums up, "In Basalla's Eurocentric model, dynamism belongs to an (improbably) homogenous West, leaving the rest of the world to participate only passively in the process of diffusion, unable to make any original contribution of its own or even to negotiate with an ascendant Western science" (12). Dhruv Raina has tried to upturn this notion by suggesting that the ideology of science has been "actively redefined" by the "recipient culture": the receiving culture "subverts, contaminates, and reorganises the ideology of science as introduced by Europe" (cited in Arnold, 13; also see his "Introduction," in Raina and Habib, 1–15). The problem with such a counterargument, which Arnold does not notice, is that in accepting the originary Europe and the recipient India, all that Raina and Habib do is to give more agency to the recipient, reducing the power of the diffuser. Their model of scientific production and reception remains not only diffusionist and Eurocentric, but also dualistic.

Arnold asks the fundamental question of what we can do if we reject this diffusionist approach. If "distinctions between centre and periphery, between 'metropolitan' and 'colonial' science, fundamentally misrepresent the way in which science evolved internationally" (13) even so how can one ignore the differential power relations between the metropolitan center and the colonial peripheries, with their still persisting hierarchies and dependencies? Science and technology "were, and surely remain, aspects of a global hegemony" (15). V. V. Raman has suggested that this hegemony will only shift if and when the non-West takes the lead in creative science and productive technology, and this can be achieved without losing one's cultural moorings.

The other point that Arnold takes up is the role of science in the creation and spread of modernity in India. As Gyan Prakash puts it, "scientific reasoning became the organizing metaphor in the discourse" (quoted in Arnold, 16) of Indian modernity, promoted not only by the colonial administration but also by an increasingly participatory native elite. Scientific evangelism, along with Christian evangelism, became the battering ram that tried to destroy the traditional cultures of India. But just as modernity was problematic to Indians, so too was modern science. As Partha Chatterjee shows, Indian nationalism attempted to mediate between the rejecting of colonial authority and the acceptance of Western modernity, thus becoming for Indians a way to affect what Sri Aurobindo called a "selective assimilation" of the West: "[Nationalism] provided a discourse . . . which, even as it challenged the colonial claim to political domination, . . . also accepted the very intellectual premises of 'modernity' on which colonial domination was based" (Chatterjee, 30).

But what this really means is that neither modern science nor modernity itself is homogenous. The diffusionist model needs to be questioned in both spheres. At the same time we need to recognize the hierarchies of power, inequality, and hegemony that mark both domains. One might argue that Indian scientists sought to forge their own brand of science as Indian intellectuals did their own brand of modernity. That this story has not been told does not mean that it cannot be told. It is just that both science and scientists are reticent if not resistant to such a proposition, undermining as it does the very fundamental, constitutive, and self-defining characteristic of science as objective, value-neutral, universal, rational, and so on. Arnold acknowledges that this was possible for Indian modernity:

> Indian scientists and intellectuals tried to construct their own brand of Indian modernity, particularly through the selective incorporation (or re-invention) of Hindu ideas and traditions, through a mix of elements, the degree of "hybridity" involved in the process, varied widely from one individual to another, even with the emergent scientific community. (17)

He doesn't take this so far as to suggest that India had the capacity and indeed demonstrated the ability to devise its own culturally unique brand of science, a sort of neo-Hindu science, if you will. I think this is one of the key questions for any serious study of science and spirituality in India. Is there a distinctly Indian science? If so, what are its defining characteristics? And what is the role of spirituality in the constitution of Indian science?

David L. Gosling, in the first book published specifically in the area of science and spirituality in India, argues that the distinctive contribution of Indian science is its holistic and integrative approach: "What has always been the most distinctive feature of Indian science is a form of integral thought, a kind of intuitive ability to hold together ideas which have elsewhere remained unrelated" (3). Quoting the work of Jagadish Chandra Bose, Gosling observes that "from the point of view of Indian scientists the progress of science in the West seemed to be a fulfilment of an important Hindu insight—the fundamental unity of all existence" (24). According to Gosling, Bose's work proceeded from the fundamental principal that "in the multiplicity of phenomena, we should never miss their underlying unity" (24).

Finally, in considering the progress of science in modern India, we cannot ignore the history of specific institutions founded for this purpose. The case of the Indian Institute of Science (IISc), Bangalore, which is one of India's leading centers of excellence in science, is illustrative. Despite the initiative and largesse of Jamsetji Tata, IISc had a very difficult start because it faced stiff opposition from the colonial authorities. As B. V. Subbarayappa shows in his painstaking study, there was both condescension and constant resistance from the colonial administration even when the entire funding for the project was to come from private philanthropy. Mooted as early as 1892, the Institute finally started to function only in 1911, seven years after Jamsetji's death. In this interim, the original plan itself went through several modifications and reinventions. The first director, an Englishman named Morris V. Travers, a professor of chemistry from University College, Bristol, was hired at a salary probably higher than what he earned in Britain. The colonial hangover and interference continued till way past the tenure of the first Indian director, C. V. Raman. Here, the rise of modern science in India clearly needed the idea and later the political fact of an independent India to support it.

The last sixty years since India's independence, we see both science and spirituality flourishing, even if in different ways. Paradoxically, they may be considered farther apart than they were during the freedom movement. It would appear that nationalism, that great unitive and cohering force, was the glue that brought them together during the heady decades of the late nineteenth and early twentieth centuries. Both spiritualists and scientists wanted India to be free and saw it as their patriotic duty to contribute and collaborate in the larger project of nation building. In today's India we see a renewed need for a serious dialogue between these two domains that are more or less seen as separate even if not incommensurable.

The challenge, as I see it, is to be able to map what I would call the *episteme* of modern India. The ancient Greeks used the term episteme in contrast to *techne*, which was largely considered inferior. In classical Indian thought too, *kala*—skill, art, craft, or technique—was distinguished from *vidya*—knowledge, understanding, insight—which could actually liberate the practitioner. Hence, *sa vidya ya vimuktaye*—that is knowledge that liberates; as opposed to that which only helps one earn a livelihood. Though episteme was often simply translated as knowledge, Michel Foucault in *The Order of Things* considered it as the very ground of knowledge or the conditions that make the creation of knowledge possible. These conditions could be a set of codes that authorize certain discourses and disallow others, or they could be a set of paradigms in the Kuhnian sense, but essentially they cut across disciplines and are not confined merely to science. I would not like to invest the episteme with either the esoteric, almost mystic diffuseness of Foucault nor the confining, almost overdeterministic power that Kuhn invests in the idea of the paradigm. To me, *episteme* suggests a set of governing ideas that though not necessarily *a priori*, nevertheless undergird and influence knowledge production in a certain geocultural space during a certain epoch.

Both the long nineteenth century in India and the relatively short twentieth century that followed it do present the possibilities of or the workings out of a distinctive Indian episteme. At least this is the hypothesis. Through the clash of contending ideas, some unique modes of synthesis and innovation took place in India. If we were to simplify the constituents of this clash, we might present it as two intersecting axes:

Figure 3.1. The Clash of Contending Ideas

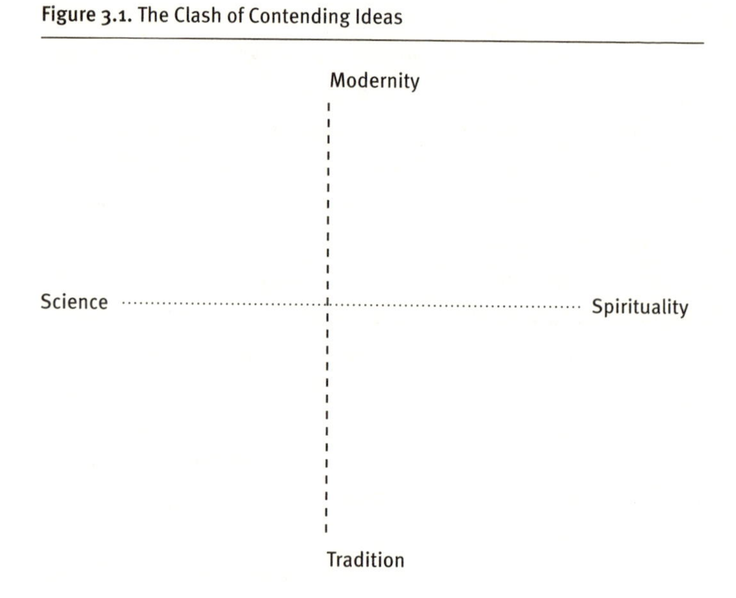


In which of the four boxes was India to fit? Should it remain traditional and spiritual? Or scientific and modern? Or, more properly, a combination of tradition and modernity *and* of science and spirituality? My theory is that India certainly wanted to become modern; the desire for modernity is still very strong. Similarly, India's welcoming of science is almost unprecedented in any "third world." Starting late, it has turned to science with an appetite that only suggests that it wanted to make up for lost time. Yet both Indian science and Indian modernity are not built upon a complete rejection of their "Others"—spirituality and tradition—respectively. Rather, there is an ambivalence, even illicit attraction for this "Other," normally to be discarded in the process of "growing up." Kant's cry of *sapare aude* ("dare to know") almost served as the motto of the European Enlightenment; the Indian modification of it might read, "dare to know anew, but do not forget the old."

Even hard-core and uncompromising children of the Enlightenment such as Karl Marx found India a puzzle. Conceding to this subcontinent a trajectory that almost escaped the universal logic of history (with a capital "H"), Marx coined the much-debated term *Asiatic mode of pro-*

*duction* to characterize the Indian economy. Similarly, when historians, social anthropologists, literary critics, and cultural theorists have tried to define Indian modernity, they have often used terms that suggest a certain complication or contamination. It is as if India inflects modernity in such a manner as to change its inner logic. The project of modernity, defined in terms of the march of triumphant rationality and scientific progress, does not quite run aground in India, but becomes deeply "polluted" with something else. In my own writings, I have called India's rite of passage neither modern nor antimodern, but radically *a*modern or nonmodern. Others have used works like "archaic modernity," "fugitive modernity," "critical traditionalism," and "alternative modernity" to try to understand what happened in India.

The foregoing discussion suggests that the interrelationship between science and spirituality in India is a historically evolving one. In its earlier phase in the late nineteenth century, the relationship was possibly closer, partly because both science and spirituality in India contributed to the creation of Indian modernity and what is perhaps more significant is that both also became central to the process of consolidation of Indian nationalism. However, the relationship between the two has changed in more recent times to one of the independence of the two domains. If we revert to Ian Barbour's classic formulation of a fourfold typology of conflict, independence, dialogue, and integration, we notice that in India all four types of relationships have been present both in the past and in the present, but that conflict has never been predominant unlike in the West. Moreover, while a number of practicing scientists stress the independence of the two domains, a number of spiritual leaders have advocated both dialogue and integration. Indeed, His Holiness, the Dalai Lama, has often stressed the need to dialogue. Dialogue, of course, is the precondition to a possible integration.

Despite the shift from a scientific universalism during high modernity to a sort of cultural relativism engendered by postmodernist philosophy, science in the sense of a well-defined, commonly accepted, multicultural enterprise continues to inform and determine constructions of truth and reality in our contemporary world. While tradition may be the repository of values, beliefs, and ways of relating to each other and the world, we need not, as V. V. Raman suggests in the conclusion to *Glimpses of Indian Science*, cling to premodern explanatory models just because they are culturally significant. To do so would make us cultural fundamentalists if not outright reactionaries. On the other hand, dogmatic scientific materialism may also result in the closure of the mind and of the possibilities of experiencing and explaining phenomena. We will have to admit that there are areas of knowledge and truth that are simply outside the parameters of science as it is understood today.

Those who believe in the evolutionary possibilities of human futures aver that we are on the brink of a global renaissance that requires the integration of not just Western and Eastern cultural and civilizational resources, but the coming together of science and spirituality. On the ground such a convergence seems as yet only a distant dream. Yet the unexplored possibilities of dialogue between these two domains afford us challenges and opportunities hitherto unexplored. India, I believe, has a crucial role to play in such a dialogue, positioned as it is. Custodian of an ancient civilization that is also the home of unique experiments in a plethora of spiritual endeavours, it is at the same time positioning itself as one of the leaders in the IT revolution that is sweeping across the globe. This makes India not just a fertile ground for such an enquiry but the bearer of a special responsibility toward the future of such a dialogue.

## Bibliography


Arnold, David. *Science, Technology, and Medicine in Colonial India*. Cambridge: Cambridge University Press, 2000.

Baber, Zaheer. *The Science of Empire: Scientific Knowledge, Civilization, and Colonial Rule in India*. Albany, NY: SUNY Press, 1996.

Barbour, Ian G. *When Science Meets Religion*. San Francisco: Harper San Francisco, 2000.

Basalla, George, ed. *The Rise of Modern Science: External or Internal Factors?* Lexington, MA: D. J. Heath, 1968.

———. "The Spread of Western Science." *Science* 156 (1967): 611–22.

Chatterjee, Partha. *Nationalist Thought and the Colonial World: A Derivative Discourse?* London: Zed Books, 1986.

Foucault, Michael. *The Order of Things: An Archaeology of the Human Sciences*. New York: Pantheon Books, 1970.

Gosling, David L. *Science and Religion in India*. Madras: Christian Literature Society, 1976.

Nandy, Ashis. *Science, Hegemony and Violence: A Requiem for Modernity.* New Delhi: Oxford University Press, 1988.

Prakash, Gyan. *Another Reason: Science and the Imagination of Modern India.* New Delhi: Oxford University Press, 2000.

Raina, Dhruv, and Irfan Habib, eds. *Situating the History of Science.* New Delhi: Oxford University Press, 1999.

Raman, V. V. *Glimpses of Indian Science.* New Delhi: Samvad India, 2006.

Subbarayappa, B. V. *In Pursuit of Excellence: A History of the Indian Institute of Science.* New Delhi: Tata McGraw Hill, 1992.

## Editor's Introduction to

# *Kokoro* [Mind-Heart-Spirit]: Affirming Science and Religion in the Japanese Context

**Paul Swanson,** Nanzan Institute for
Religion and Culture, Japan

Dr. Paul Swanson, director of the Nanzan Institute for Religion and Culture, presents the first results of a novel and productive exploration of science and spirituality contextualized by a single Japanese word: *kokoro*. A "comprehensive concept," *kokoro* is a touchstone for inquiry in the Japanese contexts and unpacking it presents unique challenges and rich potential. The meanings of *kokoro* include "thinking" and "feeling"; it is the "center of both emotive and cognitive sensitivity." Swanson and his research group begin a major scholarly undertaking by trying to effectively analyze this concept from a Western perspective while maintaining its unitary nature and defying the reductionist tendency to view it as complex—itself further separable into pieces and subunits of meaning.

Swanson's team ground their work in a long intellectual history embracing mutual dependence and inseparability. Through personal anecdotes, as well as literary and scholarly references, they make the case that a nonunitary/nondual reciprocity underlies Japanese thought. From the Buddhist precept that the real and the experienced are neither one nor two, to the poetic contradiction that giving "free reign to your desires, you become uncomfortably confined," Swanson softens the ground under a linear march toward combining science and religion in a reductionist typography. But he holds out the hope that a bridging concept like *kokoro* will offer a safer route and proceeds to offer examples from his team's research on applications of that concept in science and technology.

In robotics and climate modeling, they wrestle with the essential question of intellectual capacity. Is it enough to solve problems? Clearly a numerical computer predicts the future but computer simulation lacks the element of desire fulfillment—it seems that it cannot disturb the *kokoro.* It would be this very sort of disturbance that might prove the existence of *kokoro* in a sentient machine. If a robot can be "cultivated" that someday rebels and smacks its maker in the nose, that will be a sign of incipient independence and the emergence of a machine with *kokoro.* In medical technology, a group member suggests that a helper concept, *yasashii* (meaning tenderness, gentleness, sensitivity), offers an expansion of our view of how technology should interface with humans in light of our *kokoro* and not simply with reference to our physicality.

The human brain and mind undoubtedly include goal-directed behavior, reflexes, and instinct. But the complex of these activities is difficult to analyze through mechanistic, neurological tools like brain imaging. While one of Swanson's group members turns his attention to assessing the brain's material response to religion and spirituality, another proposes that the same complex of activities is present in our nearest mammalian relatives, the primates. For Matsuzawa Tetsurō, it is evident that primates exhibit cognition and even a recognizable "artistic style," implying that their *kokoro* is far more similar to humans' than it is different.

While himself a Christian, Paul Swanson makes an important case that careful juxtaposition of scientific and spiritual concepts across the spectrum of religions and spiritual traditions can be fertile and, perhaps, even necessary. While it may ultimately be desirable to reframe the conversation (perhaps by using terms like "traditional ways"), he argues that uniquely Japanese insights offer avenues for remarkable progress in the developing relationship between science and the human spirit.

P.D.

4

# *Kokoro* [Mind-Heart-Spirit]

## Affirming Science and Religion in the Japanese Context

**Paul Swanson,** Nanzan Institute for Religion and Culture, Japan

The human mind (and heart) and how it works is one area of mystery that is still in aching need for examination through scientific inquiry. What does it mean to think (and feel)? Is the bifurcation between thinking and feeling, cognition and emotion, or mind and heart an accurate and useful distinction when considering the integrated nature of human experience? Are the familiar Western (and some distinctive English) concepts of mind, heart, spirit, will, consciousness, soul, and so forth the best way to describe and divide human experience? Or is a broader and more inclusive concept useful for understanding how humans think/feel? The Japanese term *kokoro* is such a comprehensive concept that may prove useful for considering the interrelated activity of the human mind and heart.

Thomas Kasulis, an expert in comparative cultures and philosophies and current holder of the Nanzan Chair for Interreligious Research, addresses this question as follows:

> What is *kokoro*? For starters, we can say *kokoro* is the center of both emotive and cognitive sensitivity. So, translators often render the word into English as "heart and mind." A problem with this rendering,

however, is the conjunctive "and." It might lead one to think *kokoro* is the combined function of two separate faculties, one affective and one intellectual, but this is not the case. To translate *kokoro* as "heart and mind" is like translating the Japanese word for "water" (*mizu*) as "hydrogen with a half portion of oxygen." It is not that the translation is inaccurate exactly, but rather that it misses the point, at least in any ordinary context. When requesting a glass of *mizu*, a Japanese does not think of it as a compound of two elements. Similarly, in ordinary Japanese contexts "*kokoro*" is a simple, not a compound. If we need to use a compound expression to translate *kokoro* into English, that fact tells us more about the web of English concepts than it does about the nature of *kokoro* in the Japanese worldview. In modern Western philosophizing, we have drawn such a wedge between the affective and the cognitive that we too easily slip into believing the universality of the bifurcation. Hence, we assume that *kokoro* must have a dual rather than singular function and we translate it as such. To sum up: the "heart and mind" translation hides as much as it discloses. We think we know what *kokoro* means only by occluding its most threatening suggestion, namely, that our modern Western bifurcation between emotion and cognition may be at best limited and at worst simply wrong.[1]

Over the past couple of years, in the course of the first-stage Japanese GPSS project of conducting discussions, colloquia, and a symposium on the theme "Science—Kokoro—Religion," the Japanese notion of *kokoro* (mind/heart/spirit) has served well as an "operating concept," bridge, or focus to speak of matters of the mind, heart, and spirit.[2] Since *kokoro* is a concept that includes both "mind" and "heart," it serves as a way to address bifurcated concepts such as "mind-and-heart," "reasoning and emotion," "thinking and feeling," and, by extension, "science and religion," as interrelated and mutually dependent rather than independent and separate. In Buddhist terms, one can say that these bifurcated notions are "neither the same nor separate/distinct," "neither one nor two," or even "neither dual nor nondual, and yet both dual and nondual." T'ien-t'ai Chih-i, the founder of T'ien-t'ai Buddhism and whose influence on East Asian Buddhism and culture is akin to that of St. Thomas Aquinas in the West, spoke of how the "mundane" (everyday, conventional experiences) and "real" (the way things really are) are "neither one nor two," "neither [completely] different nor [totally] the same," "nondual yet not distinct," "neither merged nor scattered."[3] This idea of nonmonolithic nondualism, in which things retain their distinctiveness while maintaining identity or connectiveness with others or with the whole, is an important aspect of the Japanese worldview.[4]

"Nonduality," of course, is not a uniquely Asian or Buddhist concept—Jacob Needleman, for example, speaks of science and spirituality as an "organic unity," a "reciprocal relationship among separate but interdependent entities"[5]—but it is nonetheless a key perspective for the Japanese context. It is a tendency to perceive the relationships between varied phenomena rather than to focus on their individual, independent essences. Perhaps the same can be said of science and religion, "heart" and "mind," and that this nondualistic or interrelational Japanese perspective, and specifically the comprehensive concept of "*kokoro,*" can provide a new perspective for the science-religion dialogue?

Again, this idea of the interrelatedness of mind and heart is not uniquely Japanese. Recently I was having dinner with Dr. Keel Hee-sung, a Korean theologian and scholar of religions, and the conversation turned to a certain person with very strong beliefs and commitments. To make his point, Dr. Keel put his hand over his heart/chest and said, "He really believes this with all his mind." I noted that if a "Westerner" had made that comment, he/she would have pointed at his/her head and said, ". . . with all his mind." This led to a discussion of "*kokoro,*" and Dr. Keel confirmed that the Korean language uses an equivalent of "*kokoro*" with the same comprehensive meaning.

To give some examples from the popular press, a special issue of *Newsweek* magazine (International edition, October 17, 2005) on "Stress and Your Heart" introduced recent trends in the medical community that increasingly recognize the interrelationship and mutual influences of mental, emotional, and physical states (a fact that seems to me rather obvious), closing with the statement "the heart does not beat in isolation, nor does the mind brood alone" (35). An essay in the *International Herald Tribune* (August 25, 2005, reprinted from *The New York Times*) on "A Brain in the Head, and One in the Gut: Scientists Study Connection between Digestive and Psychiatric Problems," explains the new field known as "neurogastroenterology" and the recent discussion over the "second brain" in the gut known as the "enteric nervous system." It appears that due to the heavy concentration of nerves in the human digestive system, we "think" (or "feel" or at least "react") directly through the enteric nervous system in our bellies without consciously "thinking" and analyzing the situation first in the brain in the head. Hence, it is not so accurate to say that we "think" with our heads (brain) and "feel" with our gut, but that the two functions are inextricably part of an integrated nervous system that guides human behavior.

In an example from Japanese literature, the novelist Natsume Soseki (1867–1916) opens his novel *Kusamakura* ("Grass on the Wayside" or "The Three Cornered World") with one of the most famous passages in modern Japanese literature:

> Walking up a mountain track, I fell to thinking. Approach everything rationally, and you become harsh. Pole along in the stream of emotions, and you will be swept away by the current. Give free rein to your desires, and you become uncomfortably confined. It is not a very agreeable place to live, this world of ours.[6]

In other words, thoughts, feelings, desires (will) are all interrelated aspects of what it means to be human, and we would be wise to take all of them, and their interrelationship, into account in order to understand human experience.

### Reflections on First-round GPSS Activities

The use of *kokoro* as a bridge between science and religion allowed Japanese scientists to explore and discuss "spiritual" matters in a way that they would not be free to do so in their usual academic environment. Some of the most sophisticated scientific work related to these questions—in areas such as brain science, robotics, simulation science, primatology, medical technology, and so forth—is being conducted in Japan by Japanese and international scientists. Many of the best Japanese scientists in these areas participated in our first round of fourteen colloquia and a final symposium at Nanzan, and have shown an interest in a continued pursuit of these issues. Let me introduce some of the contents and themes of the colloquia and symposium to illustrate this point:[7]

### Satō Tetsuei and Simulation Science

Dr. Satō Tetsuei is the director of the Earth Simulator Center, currently (at least when he gave his presentation in May 2005) the largest concentration of computing power in the world, and is involved in using computers to make predictions concerning events such as earthquakes, typhoons, and global warming. In discussing simulation science, Dr. Satō pointed out that when the human brain tries to anticipate the future, it evaluates stored memories and uses various criteria to make a judgment. Thus it can be said that the ability "to think" is closely interconnected with the ability to "predict the future." But can a computer replace the human brain or mind? For a computer, the greater the computing capacity, the greater the accuracy of prediction, thus making estimations and predictions of the future into "science reality" rather than "science fiction." However, Satō points out:

> The human brain makes predictions inside brain nerve cells and determines behavior by comparison with the matters of the past, but this prediction cannot, strictly speaking, be called a prediction. For the most part it is rather related to the sphere of human desires and wants. . . . Computer simulation is not concerned with fulfilling one's desires or expectations, or some selfish ideas; it is concerned with the implementation of scientifically definite things. Therefore, it does not cause any frustration (anxiety or dissatisfaction) in the *kokoro*, and therefore it does not, in itself, require religion.
>
> The activity of the brain can easily shatter the (fragile) human *kokoro*, but simulation science and its predictions are robust. On the other hand, the human heart is rich in intuition; it possesses attributes such as illogicality, hunger for novelty, creativity, infinity and openness. Computer simulation is deterministic (closed); it lacks diversity and is an embodiment of dryness. I believe that this is the decisive difference between computers and human beings.[8]

### Matsuzawa Tetsurō and Primatology

Perhaps the most provocative of the colloquia was a presentation by Dr. Matsuzawa Tetsurō, one of the top primatologists in the world. His work with the chimpanzee Ai (and now Ai's son Ayumu) has shown how close the relationship is between chimpanzees and humans. Physically, there is only a 1.3 percent difference in the genetic DNA content between chimpanzees and humans, much less than the difference between a zebra and a horse, or between a rat and a mouse (about 4 percent). Matsuzawa has shown that chimpanzees are better at some cognitive skills (e.g., retention of short flashes of complicated data) than humans; they can develop their own recognizable "artistic style" through a series of paintings, and so forth, indicating that the human mind (and heart) is not unique in the animal world, at least in these respects.[9] Matsuzawa's stated goal in working with chimpanzees is "the study of *kokoro*": that it is possible to study the evolution of the human mind-and-heart by getting to know chimpanzees, and that this will shed light on what it means for humans to think and feel. The implications of this research for religion are stunning, and present strong challenges to an anthropocentric religious worldview.

### Hashimoto Shūji and Robotics

Famous for his stated goal of creating a robot with a *kokoro*, or a "sentient machine," Hashimoto Shūji of Waseda University has shifted his thinking from "building" or "creating" to "cultivating/growing/developing" a sentient machine:[10]

> My dream is the creation of various kinds of robots with self-reproductive functions and with a will to live. I want to rear robots by putting them into a skillfully arranged environment and looking out for them like a shepherd, or like a nurse caring for a baby, not letting it go too far, or pulling it back up if it falls into a drain. Then I will wait for a robot that will develop in the course of several generations and be able to discuss with us issues such as "what is a living being?" and "what is *kokoro*?" If in the process a robot rebels and hits me, with my nose bleeding I would probably rejoice in my heart, thinking, "Finally, I did it. We've almost made it!" This is because a period of rebellion naturally precedes independence.[11]

Hashimoto is also critical of Isaac Asimov's famous "Three Laws of Robotics" as too anthropocentric.

> As we ponder the society of the near future, we realize that the difference between humans and robots will become vague and the concept of "human" existence underlying Asimov's Three Laws of Robotics will become dubious. In the field of high-technology medicine, experiments are currently conducted on the production of all kinds of artificial organs, and human robotization pro-

gresses. A "happy" human brain does not yet exist. However, chemicals influencing memory and mental activity are already partially used. In addition, for the treatment of vision or hearing-related illnesses, there are surgical methods that involve the direct connection to nerves. Perhaps artificial organs with a direct connection to the brain will appear soon. As robots are getting closer to humans and humans are getting closer to mechanisms, Asimov's Three Laws of Robotics will become basically meaningless. I believe that scientific technology—not limited to robots—must elaborate a new philosophy based on the goals of humanism that will tackle questions such as "what is human?" and "what does it mean to live happily?"[12]

### Tanaka Keiji and Brain Science

Dr. Tanaka Keiji, group director at the Japanese government-sponsored RIKEN Brain Science Institute, also addresses the question of "what does it mean to think and feel" from the perspective of his very technical research on the neurological workings of the brain through brain imaging. He writes:

> To consider the mind (*kokoro*) from the viewpoint of neuroscience, let us define it provisionally as the overall mental activity controlling one's behavior by a goal-directed approach. In the case of reflexes, innate compound movements, instinctive behavior, or habitual behavior, people are not aware of the purpose of their behavior. Goal-directed behavior is only one among many types of behavior in humans. The frontal association areas (also referred to as the prefrontal cortex) play an important role in goal-directed behavioral control. . . .
>
> The results of human brain imaging studies suggest the involvement of the medial prefrontal cortex, similar to the case of an action aimed at obtaining a primary reward. It is thus possible to analyze how the mind functions in goal-directed behavioral control, and it has been demonstrated to date that the different regions of the prefrontal cortex play important roles for goal-directed behavioral responses.[13]

Dr. Tanaka plans to pursue the spiritual or religious implications of this technical research on the brain in the future.

### Tomita Naohide and "Human-friendly" Medical Technology

Dr. Tomita Naohide of the Kyoto University International Innovation Center is concerned with the ethics of medical engineering and the importance of mental and emotional serenity (*anshin*: "a peaceful *kokoro*") in addition to the nuts and bolts of medical technology itself. He uses the Japanese concept of *yasashii* (which can be translated variously as "gentle, tender, kind, affectionate, sensitive, friendly" and so forth) to urge the development of medical technology that is "human-friendly" rather than merely "efficient," concluding that

> Inevitably it will be crucial for "yasashii technology" to create an environment completely receptive of diversity, in which we will effectively deal not only with the important factors selected from human diversity, but also gently deal with every single factor, including those considered "inefficient." In addition, recently a new methodology was proposed that allows the describing of difficult-to-describe factors in the form of a narrative. These are not pseudo-scientific methods but rather new directions that should challenge science and technology in the twenty-first century. Although, to date, scientific technology has been utilizing a methodology of "planning and control," the scientific methodology of the twenty-first century must recognize diversity, and develop a methodology geared towards "nurturing" or "developing." "Diversity" is a state of the continual interconnectedness of multiple mutually supportive factors. To develop a "yasashii" technology we must, to begin with, establish contact on a human level. A constructive dialogue will not be possible unless we create an atmosphere of mutual support and encouragement.[14]

### Thoughts on Whether or Not the Science-Religion Dialogue Is a "Western" and/or "Christian" Enterprise

A large percentage (about 50 percent) of the participants in the Nanzan colloquia and symposium were Christian, much larger than the overall percentage of Christians (estimated at between 1 percent to 2 percent) in the general Japanese population. Questions remain as to whether this was a coincidence, or a reflection of the contacts known at Nanzan and the principal investigator and science advisor? Or does it indicate that it is mainly Christians—whether they are scientists, or religious studies or philosophy academics—who are interested in these issues and a science-religion dialogue?

The most "standard" or "typical" science-religion presentation (given my limited exposure to such meetings in the West) was the last session of the symposium, which consisted of two papers (by Drs. Sanda Ichirō and Yamamoto Sukeyasu) on physics and religion—both presenters were physicists, both Catholic Christians, both having spent a long time studying and teaching in the United States, and both speaking of their struggle to reconcile their Christian faith with their identity and knowledge as physicists. Their presentations provoked two opposite reactions: a positive reaction in that their "struggle" to reconcile their religious beliefs and their scientific research was perceived as "new, fresh, or different" in the Japanese context; a negative reaction in that such an attempt was seen to be "foreign" to Japanese ways of thinking (too "Western"); and that most Japanese would not be able to empathize with or understand such an attempt.

There was a suggestion that instead of "science and religion," it would be better in the Japanese (or even broader "Eastern") context to speak rather in terms of "modern technology and traditional ways." In any case, this discussion underlined the importance of considering the science-religion dialogue from a different approach, and indicated that the notion of *kokoro* is a promising one for fruitful discussions.

## Future Goals

The first round of Nanzan GPSS colloquia and symposium suggests that *kokoro* is indeed a useful operating concept for discussing the interrelatedness of science and religion in Japan. Although terminology and focus points may differ from the preceding dominant discussion of science and religion in the West, the discussion in Japan promises new insights and different approaches. The subject of "mind" (and "heart") is the focus of some of the most advanced scientific inquiry in the world, often led by Japanese scientists. Their insights, and the conceptualizations that flow from Japanese terminology and cultural assumptions, are worthy of attention and should be recognized as important contributions to human understanding.

## Notes

1. From an essay by Thomas Kasulis on "Cultivating the Mindful Heart: What We May Learn from the Japanese Philosophy of *Kokoro*." For the full essay, see the Nanzan GPSS homepage under www.nanzan-u.ac.jp/SHUBUNKEN/jp/Purojekuto/GPSS/GPSS.htm, "Reference Materials." This site also contains an essay by Jean-Noël Robert on "Some Reflections on the Meanings of Kokoro as Exemplified in Japanese Buddhist Poetry: An Instance of Hieroglossic Interaction."
2. See my essay on "Science—Spirit—Religion: Reflections on Science and Spirituality in the Japanese Context" (prepared for the first-stage GPSS project, published in the *Bulletin of the Nanzan Institute for Religion and Culture* 29 [2005]: 20–26), for thoughts on the difficulty of discussing religion in the Japanese context, and the usefulness of "*kokoro*" in this situation.
3. See, for example, the *Mo-ho chih-kuan* 摩訶止観, one of the most influential East Asian treatises on Buddhist practice and theory, Taishō Buddhist canon, vol. 46.34c16ff.
4. One caveat: I am not making the ethnocentric claim that concepts such as *kokoro* are "uniquely Japanese" and cannot be understood by non-Japanese. Terminology and concepts in any language carry their own nuances and application so that no word has its exact equivalent in another language. Nevertheless terms such as *kokoro* can be explained and understood in other language contexts, just as originally Western terms, such as "religion," can be meaningfully applied in Japan.
5. Jacob Needleman, *A Sense of the Cosmos: The Encounter of Modern Science and Ancient Truth* (New York: Arkana, 1988), xiv.
6. For a translation see Natsume Soseki, *The Three Cornered World*, trans. Alan Turner (Tokyo: Tuttle, 1965; repr., Tokyo: Regnery Publishing, 1989).
7. A full list of the Nanzan GPSS colloquia, and the contents of the symposium, along with English translations of the papers presented at the symposium, are available on the Nanzan Institute homepage at: www.nanzan-u.ac.jp/SHUBUNKEN/jp/Purojekuto/GPSS/GPSS.htm.
8. Satō Tetsuei, "Simulation Culture." www.nanzan-u.ac.jp/SHUBUNKEN/jp/Purojekuto/GPSS/GPSS.htm.
9. See, for example, Matsuzawa Tetsurō, Masaki Tomonaga, and Masayuki Tanaka, eds., *Cognitive Development in Chimpanzees* (Tokyo: Springer, 2006).
10. Hashimoto's comments are reminiscent of the arguments made recently by Ray Kurzweil and his claim that computers/robots and humans will merge to develop into super intelligent machines; see, e.g., his *The Age of Spiritual Machines: When Computers Exceed Human Intelligence* (New York: Penguin Books, 1999). Kurzweil speaks of a new paradigm "called an evolutionary (sometimes called genetic) algorithm. The system designers don't directly program a solution; they let one emerge through an iterative process of simulated competition and improvement" (81).

11. Hashimoto Shūji, "A New Relationship between Humans and Machines: Is It Possible to Create Machines With Heart/Kokoro?" www.nanzan-u.ac.jp/SHUBUNKEN/jp/Purojekuto/GPSS/GPSS.htm, pp. 8–9. This is a rather different attitude for a "creator" with regard to his "creation," compared to the Christian story in which rebellion against the creator is the mark of sinfulness.
12. Ibid., 8.
13. Tanaka Keiji, "Mind (Kokoro), Goal-directed Behavior, and Prefrontal Association Areas," www.nanzan-u.ac.jp/SHUBUNKEN/jp/Purojekuto/GPSS/GPSS.htm, pp. 1–10.
14. Tomita Naohide, "Diversity and 'Yasashisa' in Medical Engineering: A Call for 'Human-Friendly' Technology," www.nanzan-u.ac.jp/SHUBUNKEN/jp/Purojekuto/GPSS/GPSS.htm, pp. 3–4.

## Bibliography


Hashimoto Shūji. "A New Relationship between Humans and Machines: Is It Possible to Create Machines With Heart/Kokoro?" www.nanzan-u.ac.jp/SHUBUNKEN/jp/Purojekuto/GPSS/GPSS.htm.

Kasulis, Thomas. "Cultivating the Mindful Heart: What We May Learn from the Japanese Philosophy of *Kokoro*." www.nanzan-u.ac.jp/SHUBUNKEN/jp/Purojekuto/GPSS/GPSS.htm, "Reference Materials."

Kurzweil, Ray. *The Age of Spiritual Machines: When Computers Exceed Human Intelligence*. New York: Penguin Books, 1999.

Matsuzawa Tetsurō, ed. *Primate Origins of Human Cognition and Behavior*. Tokyo: Springer, 2001.

———. Masaki Tomonaga, and Masayuki Tanaka, eds. *Cognitive Development in Chimpanzees*. Tokyo: Springer, 2006.

Natsume Soseki. *The Three Cornered World*. Translated by Alan Turner. Tokyo: Tuttle, 1965 (repr., Tokyo: Regnery Publishing, 1989).

Needleman, Jacob. *A Sense of the Cosmos: The Encounter of Modern Science and Ancient Truth*. New York: Arkana, 1988.

Satō Tetsuei. "Simulation Culture." www.nanzan-u.ac.jp/SHUBUNKEN/jp/Purojekuto/GPSS/GPSS.htm.

Swanson, Paul L. "Science—Spirit—Religion: Reflections on Science and Spirituality in the Japanese Context." *Bulletin of the Nanzan Institute for Religion and Culture* 29 (2005): 20–26.

Taishō Buddhist canon: *Taishō shinshū daizōkyō* 大正新脩大藏經. 100 vols. Edited by Takakusu Junjirō 高楠順次郎, et al. Tokyo: Taishō Issaikyō Kankōkai, 1924–1935.

Tanaka Keiji, "Mind (Kokoro), Goal-directed Behavior, and Prefrontal Association Areas." www.nanzan-u.ac.jp/SHUBUNKEN/jp/Purojekuto/GPSS/GPSS.htm.

Tomita Naohide, "Diversity and 'Yasashisa' in Medical Engineering: A Call for 'Human-Friendly' Technology." www.nanzan-u.ac.jp/SHUBUNKEN/jp/Purojekuto/GPSS/GPSS.htm.

## Editor's Introduction to

# Daoism and the Uncertainty Principle

**Jiang Sheng,** Shandong University, China

Daoism offers a rich and comprehensive framework within which to come to grips with the apparent contradictions between modern science and spiritual experience. Like the paradox of quantum complementarity that precludes any single logic from spanning a complete description of particle physics, and the uncertainty principle that places an upper limit on our knowledge about tiny physical objects, the Dao is an overarching worldview that embraces internal contradiction. Jiang Sheng presents an overview of the conundrums of modern physics in the context of the long and subtle traditions of Daoist thought.

Jiang, citing the pillars of Daoism, Lao Zi, and Zhuang Zi, argues that the very meaning of *Dao* is uncertainty. And this uncertainty, the Dao, "hides itself" in exactly the same cloak of unknowability that the modern physicist wraps around the concept of the void—"a union of plenum and vacuum, space and materiality." Drawing from Daoist parables, he goes on to argue that the limits of comprehensive analytic frameworks in quantum mechanics are those of perspective, that framing questions implicitly prepare the answers to be found. It behooves us, therefore, to guard against objectivist biases and embrace both the Western adjuration to "know thyself" and the Daoist "there must first be a True Man before there can be true knowledge."

In the well-known yin-yang symbol and its historical antecedents, the *Taiji* diagrams, Jiang finds an expressive pictorial tool that captures the essential problem of dynamic tension and "undulation" in our search for understanding. The no-being of yin and the being of yang are two mutually dependent elements of a dynamic complex, the thing that he refers to as "Three," "Three is not a Newtonian object, nor an objective, nor a substance. It is a *status*, an ever-lasting wave process in-between the Two" (the yin and yang). The interaction between the researcher and the material world is one such dual, and their ongoing relationship is an example of this fluctuating status. Jiang compares this idea to David Bohm's philosophy of physics and J. A. Wheeler's dynamic description of physicality.

"The Dao is the totality of all possibles and impossibles," writes Jiang Sheng. But it cannot be grasped intentionally. Like a bird, its essence exists in flight and, once captured, it loses its primary defining character, freedom. Citing the allegory of Hun-tun, Jiang asserts that the very act of drilling into the Dao destroys it. The Dao, like modern science, is a reciprocal, ongoing flux. As with the constant overturn and supercession of scientific knowledge, so too the Dao requires constant creativity. Real progress toward deep knowledge will require both spiritual and scientific exploration that humbly acknowledges our own limits and the need for constant self-reflection.

In the subtle narratives of Daoist thought and the oblique parables of its teaching, Jiang finds resonance with Wilson's consilience and Heisenberg's philosophical ruminations. With the Dao as a unifying framework, modern science and the human experience are both "clothed as with a garment," an enwrapping thought-structure that liberates the human spirit while spurring the enquiring mind.

P.D.

5

# Daoism and the Uncertainty Principle

**Jiang Sheng,** Shandong University, China

The uncertainty principle of quantum physics has posed forceful challenges to scientists and philosophers. The Dao,[1] the Oneness, the Three, and the union of known and unknown in the Dao present important parallels with the uncertainty principle and offer important ideas for the study of the universe and man himself as well.

Science as a culture is driven by the tension between man and nature. Blaise Pascal wrote in the *Pensées*, "The eternal silence of these infinite spaces frightens me."[2] Doubtless it was not the immenseness of the universe, but the limit of man's ability to know that made the mathematician—a notable representative of the Age of Reason—feel humble. Pascal understood the mind's incapacity to encompass the universe but found man's essential dignity in the attempt.

> Man is only a reed, the weakest thing in nature; but he is a thinking reed. The whole universe does not need to take up arms to crush him; a vapor, a drop of water, is enough to kill him. But if the universe were to crush him, man would still be nobler than what killed him, because he knows he is dying and the advantage the universe has over him. The universe knows nothing of this. All our dignity consists, then, in thought.[3]

In 1820 Par Pierre Simon Laplace, who saw the world as a complicated clock governed by Newton's Laws and who attempted a mathematical proof of the stability of the solar system,[4] declared that

> we ought to regard the present state of the universe as the effect of its antecedent state and as the cause of the state that is to follow. An intelligence knowing all the forces acting in nature at a given instant, as well as the momentary positions of all things in the universe, would be able to comprehend in one single formula the motions of the largest bodies as well as the lightest atoms in the world, provided that its intellect were sufficiently powerful to subject all data to analysis; to it nothing would be uncertain, the future as well as the past would be present to its eyes. The perfection that the human mind has been able to give to astronomy affords but a feeble outline of such an intelligence.[5]

But a hundred years later Werner Heisenberg published the uncertainty principle in quantum physics, declaring that man cannot exactly grasp "nature in itself," all that we may do is to observe "nature exposed to our method of questioning."

Daoism provides us with subtle tools with which to grasp the origin and the processes of reality. The idea of "Dao," along with the expression of Daoist methodology—the "ever-returning" way of thinking and grasping-in-process—offers an excellent way for modern scientists, artists, and thinkers to explore both the outer and the inner worlds.

The Dao can provide an open structure for ushering in and blending the merits of Eastern and Western civilizations. "The Dao is an empty vessel; it is used, but never filled."[6] "The relation of the Dao to the entire world is like that of the great rivers and seas to the streams from the valleys."[7] In the Dao, all different ways of knowledge find harmony with each other through the Oneness, the union of science and spirituality of different traditions.

### Quantum Theory and the Uncertainty Principle

In the past hundred years, quantum physics has posed vital challenges to classical physics and philosophy. We have been forced to revise our ideas of consciousness, reality, and speed by the revolutionary theories introduced in the 1920s by Max Planck and Werner Heisenberg.

Newtonian classical mechanics asserts that it is possible for man to precisely determine both the position and the momentum of an object at one moment; thus the future movement of that object can also be precisely predetermined. Accordingly, all relations in the universe can be precisely determined and described. Under the framework of the Newtonian system, it seemed possible that science, society, and its ever-expanding objectives could be accurately known, grasped, and correctly used.

This comfortable expectation was overthrown by quantum mechanics and its touchstone, the uncertainty principle. In an article entitled "Ueber den anschaulichen Inhalt der quantentheoretischen Kinematik und Mechanik," Heisenberg introduced the uncertainty principle, of which the core is, "The more precisely the position is determined, the less precisely the momentum is known, and conversely."[8]

In *A Brief History of Time*, Stephen Hawking asserts that

> the uncertainty principle had profound implications on the way in which we view the world. . . . We could still imagine that there is a set of laws that determine events completely for some supernatural being, and who could observe the present state of the universe without disturbing it. However, such models of the universe are not of much interest to us ordinary mortals. . . . In this theory particles no longer had separate, well-defined positions and velocities that could be observed, instead, they had a quantum state, which was a combination of position and velocity.[9]

The simple mechanics of a predictable Newtonian world have been replaced by a subtle, uncertain substrate.

The uncertainty principle and its resonance in the philosophy of knowledge is strongly reminiscent of a basis of the work of the ancient Chinese Daoist sage Lao Zi in his *Dao De Jing*, whose opening sentences are:

> The Tao that can be trodden is not the enduring and unchanging Tao. The name that can be named is not the enduring and unchanging name. [Conceived of as] having no name, it is the Originator of heaven and earth; [conceived of as] having a name, it is the Mother of all things.[10]

Another early Daoist Master Zhuang Zi writes:

The Way cannot be brought to light; its virtue cannot be forced to come.[11]

Dark and hidden, [the Way] seems not to exist and yet it is there; lush and unbounded, it possesses no form but only spirit; the ten thousand things are shepherded by it, though they do not understand it—this is what is called the Source, the Root.[12]

The Way cannot be heard; to listen for it is not as good as plugging up your ears. This is called the Great Acquisition.[13]

The essential message of these classic Daoist texts is that the Dao always hides itself in the characteristics of uncertainty; Dao means uncertainty.

According to Albert Einstein:

There is no such thing as an empty space, i.e., a space without a field. Space-time does not claim existence on its own, but only as a structural quality of the field. I wished to show that space-time is not necessarily something to which one can ascribe a separate existence, independently of actual objects of physical reality. Physical objects are not in space, but these objects are spatially extended. In this way the concept "empty space" loses its meaning.[14]

And David Bohm showed that according to our current understanding of physics, every region of space is awash with different kinds of fields composed of waves of varying lengths, while each wave always has at least some energy; "space is as real and rich with process as the matter that moves through it reaches full maturity in his ideas about the implicate sea of energy. Matter does not exist independently from the sea, from so-called empty space. It is a part of space."

Space is not empty. It is full, a plenum as opposed to a vacuum, and is the ground for the existence of everything, including ourselves. . . . Despite its apparent materiality and enormous size, the universe does not exist in and of itself but is the stepchild of something far vaster and more ineffable. More than that, it is not even a major production of this vaster something but is only a passing shadow, a mere hiccup in the greater scheme of things.[15]

Other physicists argue that "space is an illusion, so the coherence of the world must be behind and outside of space. While space may be a useful construct for certain purposes, a fundamental theory cannot be about particles moving in space. Space must only emerge as a kind of statistical or averaged description, like temperature."[16]

All these statements remind us of the ideas of *Dao De Jing*:

The great square has no corners.
The great vessel takes long to fashion.
The great note is soundless.
The great image has no form.
The Dao hides in namelessness.
It is good at giving and perfecting.[17]

Likewise, Lao Zi asserts that it is impossible for us to grasp and give an exact description of the Dao, because even the sage himself admits that he cannot define it exactly:

There was something formless and perfect
before the universe was born.
It is serene. Empty.
Solitary. Unchanging.
Infinite. Eternally present.
It is the mother of the universe.
For lack of a better name,
I force it a style name the Dao,
and call it the Great.[18]

The universe was born from the Dao, the Great, which is the mother of the universe. It gives birth to all things, and includes all things in it, but the essence of it is empty, solitary, unchanging, infinite, and ever bigger than the universe! The cosmos is a unified system. Things that seem to be separated are actually connected in fundamental ways that transcend the limitations of ordinary space and time. Lao Zi tells that everything in essence is a union of plenum and vacuum, or space and materiality.

Then what and how can we know?

According to Heisenberg, "We have to remember that what we observe is not nature in itself but nature exposed to our method of questioning."[19] He divided the objective into two layers: "nature in itself" and "nature exposed to our method of questioning." The famous Chinese tale "Little Horse Crossing River" offers a route to understanding this duality:

A little horse was asked by his mother to transmit a bag of food, but a river cut off his way. He wandered up and down and hesitated before he went to consult uncle cattle and little squirrel. Uncle cattle told him the water is not deep, and the depth is just like the length of his shank. But little squirrel warned him not to cross the river, and said that the water is very deep and a squirrel friend had just been drowned a few days ago. With no idea, little horse had to get back to ask his mother for advice. Horse mother encouraged him to try by himself. Finally little horse found that the river water was not as shallow as told by uncle cattle, nor as deep as told by squirrel.[20]

The different leg-lengths of uncle cattle, little squirrel, and little horse are just like our different methods of questioning. And with different methods of questioning, we get different knowledge of nature exposed to us. This means that whenever you arrange a different way to question the objective, you get different knowledge accordingly. Daoist Master Zhuang Zi revealed his appreciation of this subtlety in a dialogue:

No-beginning said, "Not to understand is profound; to understand is shallow. Not to understand is to be on the inside; to understand is to be on the outside."

Thereupon Grand Purity gazed up and sighed, saying, "Not to understand is to understand? To understand is not to understand? Who understand the understanding that does not understand?"

No-beginning said, "The Way cannot be heard; heard, it is not the Way. The Way cannot be seen; seen, it is not the Way. The Way cannot be described; described, it is not the Way. That which gives form to the formed is itself formless—can you understand that? There is no name that fits the Way."[21]

Different methods of questioning and observing form the basis of all knowledge. Appreciating this fact, we must pay as close attention to our inner world (where our questions are formed) as to the outer world that we wish to know. This is the reason why the famous inscription "Know Thyself" appears in the temple of Apollo in Olympus, and why Zhuang Zi says, "There must first be a True Man before there can be true knowledge."

In *An Essay on Man*, Ernst Cassirer says:

The initial steps toward man's intellectual and cultural life may be described as acts which involve a sort of mental adjustment to the immediate environment.

But as human culture progresses we very soon meet an opposite tendency of human life. From the earliest glimmering of human consciousness we find an introvert view of life accompanying and complementing this extrovert view. . . . The question of the origin of the world is inextricably interwoven with the question of the origin of man. . . . Henceforth self-knowledge is not conceived as a merely theoretical interest. It is not simply a subject of curiosity or speculation; it is declared to be the fundamental obligation of man.[22]

But how can we know our nature before we can know the objective? Jose Ortega y Gasset states:

What has taken place, . . . is the "substantial" change in the reality "human life" implied by man's passing from the belief that he must exist in a world composed only of arbitrary wills to the belief that he must exist in a world where there are "nature," invariable consistencies, identity, etc. Human life is thus not an entity that changes accidentally, rather the reverse: in it the "substance" is precisely change, which means that it cannot be thought of Eleatically as substance.[23]

Man is what has happened to him, what he has done.[24]

Man, in a word, has no nature; what he has is . . . history. Expressed differently: what nature is to things, history, *res gestae*, is to man.[25]

The history of science has shown that man has the capacity to "update" his systems of knowing. But this always happens when he is forced into a corner. We find ourselves in such a corner now. David F. Peat notes that

quantum theory forces us to see the limits of our abilities to make images, to create metaphors, and push language to its ends. As we struggle to gaze into the limits of nature we dimly begin to discern something hidden in the dark shadows. That something consists of ourselves, our minds, our language, our intellect, and our imagination, all of which have been stretched to their limits.[26]

Lao Zi's words echo this sentiment:

Existence and non-existence gives birth the one to the other.[27]

The thirty spokes unite in the one nave; but it is on the empty space, that the use of the wheel depends. Clay is fashioned into vessels; but it is on their empty hollowness, that their use depends. The door and windows are cut out to form an apartment; but it is on the empty space, that its use depends. Therefore, what has an existence serves for profitable adaptation, and what has not that for usefulness.[28]

"Existence" always indicates the other side of itself, "nonexistence"; any one of them is just a "reminder" or a "tool" of and by the other or the whole. Certainty is in fact a phenomenon (also a tool) and a reminder of uncertainty. We should break away from old-fashioned single-sided knowledge.

## Daoism: Union of the Known and the Unknown

Daoism comes from a very ancient tradition in China. Because of its emphasis on change and uncertainty in nature and society, it has long been seen as an unorthodox ideology compared to Confucianism, which rejects change (the latter insists "the way [Dao] will never change as long as Heaven is not changed"). But the fact is that in the Chinese history of spiritual life, with the transcendental ideal of gaining longevity or immortality through practicing the mysterious exercises for the Oneness with the Dao, Daoism was of cardinal significance in the lives of emperors and laypeople.

And now, *Dao De Jing* is the most widely translated and published book of all time after the Bible; it has been translated into more than twenty languages with more than five hundred versions, to which is being added at least three to five new versions per year.

Why has Daoism become so popular?

Zhuang Zi, the second most important Daoist sage, presented a famous humorous dialogue which is used to show where the Dao is to be found or "grasped":

> Where is the Dao?
>
> Master Tung-kuo asked Chuang Tzu, "This thing called the Way—where does it exist?"
>
> Chuang Tzu said, "There's no place it doesn't exist."
>
> "Come," said Master Tung-kuo, "you must be more specific!"
>
> "It is in the ant."

**Figure 5.1.** *Dao De Jing* and *Taiji* (Supreme Ultimate): The Evolution of the Universe from the Dao to All Things

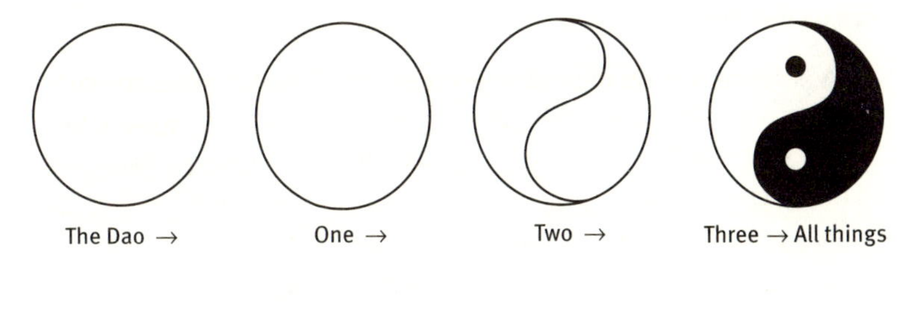


"As low a thing as that?"
"It is in the panic grass."
"But that's lower still!"
"It is in the tiles and shards."
"How can it be so low?"
"It is in the piss and shit."[29]

The Dao is the property of all things. It exists everywhere, in everything. Each observer has an inherent ability to find out Dao, thus Dao demonstrates itself particularly in each observer's view. In traditional Chinese culture, "the Dao" is the ultimate expression of all wisdom. What cannot be seen is called indistinguishable. What cannot be heard is called indistinct.

What cannot be touched is called indefinite. These three can't be comprehended, So they're confused and considered one. Its surface is not bright. Its depths are not obscured. Dimly seen it can't be named, so returns to the insubstantial. This is the shapeless shape, the form without substance. This is called blurred and shadowy. Approach it you can't see its face. Follow it you can't see its back.[30]

How shadowy, how indistinct! Within it is the form. How shadowy, how indistinct! Within it is the "thing." How dim, how dark! Within it is the substance. The substance is perfectly real, Within, it can be tested. From present to ancient times, Its name was never lost. So we can investigate the origins of all. How do I know the origins of all are that? By means of this.[31]

Generation by generation, Chinese scholars transferred the tradition of looking upward to observe 'celestial characters' (literally translated from Chinese words that mean astronomical phenomenon), looking downward to examine 'earth grains' (literally translated from Chinese words that mean geographical phenomenon) which is recorded in *the I-Ching* (*the Yi Jing*, or *the Book of Changes*), and expressed excellently by Lao Zi in *Dao De Jing*: "The Tao gives birth to One. One gives birth to Two. Two gives birth to Three. Three gives birth to all things. All things have their female and male contents in themselves; in-between them vigorous vitality emerges by which harmony is achieved."[32]

More discussions on the dynamics of the changes of all things of the universe are found in Lao Zi's statements:"Being and non-being create each other. Difficult and easy support each other. Long and short define each other. High and low depend on each other. Before and after follow each other. This lasts forever."[33]

As for the characteristic of the two contents (yin and yang) in *Taiji* operation progress, Lao Zi states: "The Dao of heaven is like stringing a bow. It depresses the high, and raises the low. It takes from excess, and gives to the lacking. It is heaven's Dao to take from excess, and give to the lacking."[34]

Lao Zi has clearly expressed the tension in everything of the universe by his vivid "stringing a bow," which is the root of the dynamics of all changes and movements, including those of nature, society, and human life. This tradition gave birth to a great idea that was expressed by the so-called Diagram of the Supreme Ultimate (*Taijitu*). An early image and interpretation of it can be found in Zhou Dunyi's works. Zhou Dunyi (or Chou Tun-I, 1017–73) was the forerunner of Neo-Confucianism and founder of Daoxue in the Song Dynasty (960–1279). Zhu Xi (or Chu His, 1130–1200), a Song Dynasty Confucian scholar who also studied Daoism and became one of the most significant Neo-Confucians in Chinese history, wrote that Zhou's diagram of the Supreme Ultimate originated from the Daoist scripture *A Comparative Study of the Zhou (dynasty) Book of Changes* by Wei Bo-Yang.

The Diagram of Supreme Ultimate excellently expressed the essence of chapter 42 of *Dao De Jing*: the round circle indicates the One born in the Dao; all changes of all dimensions are included and revealed in it.

But how, then, are the changes understood by Lao Zi? The great sage

Figure 5.2. Versions of the *Taiji* Diagram

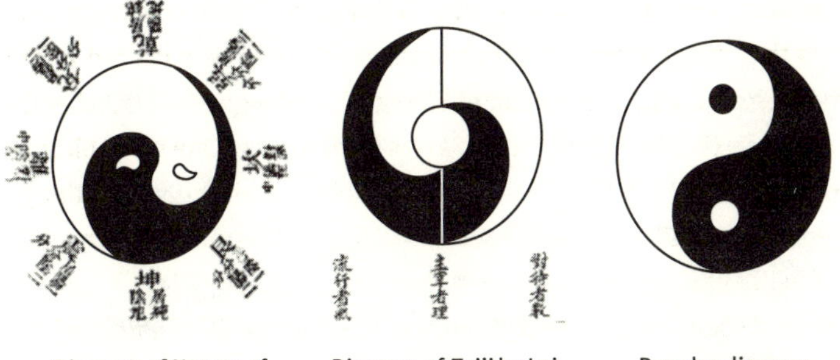

Diagram of Nature of Heaven and Earth

Diagram of *Taiji* by Lai Zhide (1535–1604)

Popular diagram of *Taiji*

used an implied number: "Three." Thus, in Daoism, "Three" becomes an ever-vigorous and ever-lively PROCESS, not the outcome. To Lao Zi, when he says "Two gives birth to Three," he indicates that from here, all things will be born or eliminated through the mutual intercourse of the female and male (in Chinese terms, yin and yang); no being (yin) means to be brought into being (yang); being means to be taken back into no being. His "Three" will never be a solid status, a describable "existence," but as an ever-vibrating or ever-undulating status, it will never stop. The life of all things lies here. This is why vigorous vitality emerges in all things; just like, for example, vapor will be generated in-between fire and water. It is here that Lao Zi finds the way to "immortality." Keep your balance through "keeping to the original Oneness,"[35] and then you will grasp the root of life.

So "Three" is a vibrating or waving status, an ever-lasting process of "stringing a bow," in which yin and yang embrace and eat and beget each other in Oneness, and in-between them vigorous vitality emerges.[36] The *Taiji* diagram can be taken as an abstract expression of the rules or relations in the universe. Almost all photos of spiral galaxies show surprisingly similarities to the diagram of Supreme Ultimate; the way of the universe is like the status of "Three" that Lao Zi expressed in *Dao De Jing*.

According to Lao Zi, if one wants to know what is yin and yang, he could only try to understand in "Three"—the process, or the status for yin and yang to exist; because yin and yang never exist separately, one of them is ever rooted in the other. But the problem is, "Three" is not a Newtonian object nor a substance but a status, an everlasting waving process in-between the Two; it is the way for the Two to exist. In "Three," the Two are mutually dependent, mutually begetting and mutually replacing. "Three" is an "ever-returning" status, and the way of thinking and grasping-in-process as well for us. Because "returning is the motion of the Tao, yielding is the way of the Tao."[37] Here one can find that "Three" exists in a wave, produced by the Two. This strongly reminds us of modern cosmology, "the quantum state of a spatially closed universe can be described by a wave function."[38]

Lao Zi's idea of "keeping to the center"[39] has the same meaning as the idea of understanding "Three." The "center" is not necessarily understood to be *the* exact solid and central point of the system but *a* dynamic central point. In certain circumstances, this point might even come close to the edge of the system, as in the elliptical path of planets. All of these are called "natural" in Daoism, which indicates the "vitality" of all things (the normal and the abnormal) in the universe.

By "keeping to the center," one can finally achieve a double-denying, nonpersistence, and thoroughly purified status called the True Knowledge and become the True Man (Man in union with the Dao). This is the realm of persistence-free. The famous Daoist Wang Xuanlan writes:

> Mysterious Virtue has two meanings: one is the Noumenon Mysterious Virtue, the other is Practice Mysterious Virtue. . . . Noumenon Mysterious Virtue is so called because the subtle ultimate noumenon is true while void; The Practice Mysterious Virtue is so called because through practice conforming to the Dao, one can cultivate himself into the Mysterious Union with the One.[40]

Lao Zi knows that it is impossible for man to grasp the real objective world by images, sounds, and feelings; they are merely a part of those ("Three") existing and floating in-between the objective and our sensing system:

> We look at it, and we do not see it, and we name it "the Equable." We listen to it, and we do not hear it, and we name it "the Inaudible." We try to grasp it,

and do not get hold of it, and we name it "the Subtle." With these three qualities, it cannot be made the subject of description; and hence we blend them together and obtain The One.[41]

Who can of Tao the nature tell? Our sight it flies, our touch as well. Eluding sight, eluding touch, the forms of things all in it crouch; eluding touch, eluding sight, there are their semblances, all right. Profound it is, dark and obscure; Things' essences all there endure. Those essences the truth enfold of what, when seen, shall then be told.[42]

Thus what you hold in hand is not what you hoped to hold. In a simple example, the bird captured by your hand is no more a bird but matter; a bird is the body and its function (i.e., flying) along with the open space and time in which it lives freely. Based on the conjugate status of the being and nonbeing, or existing and nonexisting, or seen and nonseen, known and nonknown, Lao Zi warns us against the human desire for seizing or grasping something materially, or consciously. For this Lao Zi says, "The further he goes, the less he knows. . . . For learning, he will devote himself to increase, for Dao; he will devote himself to diminish."[43] Furthermore, once we intervened in a matter, we see that what we found or grasped is not the vital original matter—only the dregs of the essence can be found. Zhuang Zi illustrated this concept in a very famous story, the Yellow Emperor went to visit Master Kuang Ch'eng for advice, saying:

"I would like to get hold of the essence of Heaven and earth and use it to aid the five grains and to nourish the common people. I would also like to control the yin and yang in order to ensure the growth of all living things. How may this be done?" Master Kuang Ch'eng said, "What you say you want to learn about pertains to the true substance of things, but what you say you want to control pertains to things in their divided state. Ever since you began to govern the world, rain falls before the cloud vapors have even gathered, the plants and trees shed their leaves before they have even turned yellow, and the light of the sun and moon grows more and more sickly. Shallow and vapid, with the mind of a prattling knave—what good would it do to tell you about the Perfect Way!"

The Yellow Emperor withdrew, gave up his throne, built a solitary hut, spread a mat of white rushes, and lived for three months in rezenfold the spirit in quietude and the body will right itself. Be still, be pure, do not labor your body, do not churn up your essence, and then you can live a long life. . . .

Heaven and earth have their controllers, the yin and yang their storehouses. You have only to take care and guard your own body; these other things will of themselves grow sturdy. As for myself, I guard this unity, abide in this harmony, and therefore I have kept myself alive for twelve hundred years, and never has my body suffered any decay."[44]

The wholeness of body and soul is the way to be almighty: to be able to grasp the world through keeping the Oneness with it (this is just like in quantum physics where the observer and the observed become a whole body), and to be able to earn longevity. Zhuang Zi also enlightens us with an inspiration that the development of science and technology is not supposed to be separately from that of spirituality of humankind. Only in this way can we dream to "get hold of the essence of Heaven and earth."

This also gives us a holographic view of things in the universe. There exists a strong similarity here to the methodology in cosmology. David Bohm says that ultimately we have to understand the entire universe as "a single undivided whole."[45] Instead of separating the universe into living and nonliving things, Bohm sees animate and inanimate matter as inseparably interwoven with the life force that is present throughout the universe, and that includes not only matter but also energy and seemingly empty space. Life is dynamically flowing through "the fabric of the entire universe."[46] And John Wheeler explains that material things are "composed of nothing but space itself, pure fluctuating space . . . that is changing, dynamic, altering from moment to moment"; "Of course, what space itself is built out of is the next question. . . . The stage on which the space of the universe moves is certainly not space itself. . . . The arena must be larger: superspace . . . (which is endowed) with an infinite number of dimensions."[47] What is the "superspace" then? For inspiration, maybe we can compare it to the Dao.

The Dao is the totality of all possibles and all impossibles, all knowns and all unknowns. The Dao presents itself everywhere. As the source of our existence, the Dao, or the mother of the universe, is forever beyond the ability of our limited mentalities to capture conceptually.

The Way cannot be thought of as being, nor can it be thought of as nonbeing. In calling it the Way, we are only adopting a temporary expedient. . . . The perfection of the Way and things—neither words nor silence are worthy of expressing it. Not to talk, not to be silent—this is the highest form of debate.[48]

In *Zhuang Zi*'s amazing Daoist allegory *Hun-dun* (or *Hun-tun*, the initial status of the universe, chaos), he indicates that our normal knowledge may only kill our opportunity to achieve ultimate knowledge:

The emperor of the South Sea was called Shu [Brief], the emperor of the North Sea was called Hu [Sudden], and the emperor of the central region was called Hun-tun [Chaos]. Shu and Hu from time to time came together for a meeting in the territory of Hun-tun, and Hun-tun treated them very generously. Shu and Hu discussed how they could repay his kindness. "All men," they said, "have seven openings so they can see, hear, eat, and breathe. But Hun-tun alone doesn't have any. Let's try boring him some!"

Every day they bored another hole, and on the seventh day Hun-tun died.[49]

In chapter 2 of the same work, he presents another seven-stage process, this time for the beginning of the universe.

There is a beginning. There is a not yet beginning to be a beginning. There is a not yet beginning to be a not yet beginning to be a beginning. There is being. There is nonbeing. There is a not yet beginning to be nonbeing. There is a not yet beginning to be a not yet beginning to be nonbeing.[50]

The Way (Dao) has never known boundaries.[51]

A few hundred years later, the great Daoist encyclopedia *Huainan Zi*, explained Zhuang Zi's position.[52]

How could it be able for one to know then?

According to Zhuang Zi: "Knowledge must wait for something before it can be applicable, and that which it waits for is never certain. . . . There must first be a True Man before there can be true knowledge."[53] What is True Man?

True Man is a man of the whole. True Man is a totality of the subject and the object. Here, he is in Oneness with the Dao, where "the known and the unknown combine in Oneness." The wisdom of nonadherence. How great it is! And this is the fine essence of Daoism, which is often wrongly translated as "nonaction."

Duke Huan was in his hall reading a book. The wheelwright, who was in the yard below chiseling a wheel, laid down his mallet and chisel, stepped up into the hall, and said to Duke Huan, "This book your Grace is reading, may I venture to ask whose words are in it?"

"The words of the sages," said the duke.

"Are the sages still alive?"

"Dead long ago," said the duke.

"In that case, what you are reading there is nothing but the chaff and dregs of the men of old."

"Since when does a wheelwright have permission to comment on the books I read?" said Duke Huan. "If you have some explanation, well and good. If not, it's your life."

Wheelwright P'ien said, "I look at it from the point of view of my own work. When I chisel a wheel, if the blows of the mallet are too gentle, the chisel slides and won't take hold. But if they're too hard, it bites in and won't budge. Not too gentle, not too hard—you can get it in your hand and feel it in your mind. You can't put it into words, and yet there's a knack to it somehow. I can't teach it to my son, and he can't learn it from me. So I've gone along for seventy years and at my age I'm still chiseling wheels. When the men of old died, they took with them the things that couldn't be handed down. So what you are reading there must be nothing but the chaff and dregs of the men of old."[54]

This seems in keeping with the spirit of modern science, which is based on the "ever-returning" way of thinking and innovating: we hope that what we find today will be superceded tomorrow. But the progress of human civilization is rooted not only in the human ability to innovate but also in his capacity for self-reflections. Moreover, what the Wheelwright meant was that man's ability is not only gained from learning but also, perhaps more notably, flowers from his insight. To gain a reliable "ability," we may need a marriage of the merits from both science and the spiritual domains of different traditions.

## Conclusion

The scientific-industrial era has afforded humankind great benefits, but it is also generating far more problems than it is solving. This is why I call the nineteenth to mid-twentieth centuries the "Grey Ages" and appeal for a "Re-enlightenment." The situation forces humankind to expand its horizons of knowledge to a higher and more inclusive level.

Albert Einstein said, "Problems can't be solved within the mindset that created them."[55] How can we get to such a new level? Such a new and inclusive level should be the path to our future. Heisenberg argues that

it is probably true quite generally that in the history of human thinking the most fruitful developments frequently take place at those points where two different lines of thought meet. These lines may have their roots in quite different parts of human culture, in different times of different cultural environments of different religious traditions; hence if they actually meet, that is, if they are at least so much related to each other that a real interaction can take place, then one may hope that new and interesting developments will follow.[56]

We have shown that Heisenberg's uncertainty principle, from the philosophical point of view, is very similar to Lao Zi's thought in *Dao De Jing*.

And Professor Edward O. Wilson of Harvard University holds that

it is not enough to repeat the old nostrum that all scholars, natural and social scientists and humanists alike, are animated by a common creative spirit. They are indeed creative siblings, but they lack a common language. . . . There is only one way to unite the great branches of learning and end the culture wars. It is to view the boundary between the scientific and literary cultures not as a territorial line but as a broad and mostly unexplored terrain awaiting cooperative entry from both sides. The misunderstandings arise from ignorance of the terrain, not from a fundamental difference in mentality.[57]

It is clear that Wilson's idea of "consilience" strongly resembles the Dao.

All-pervading is the Great Tao! It may be found on the left hand and on the right. All things depend on it for their production, which it gives to them, no one refusing obedience to it. When its work is accomplished, it does not claim the name of having done it. It clothes all things as with a garment, and makes no assumption of being their lord;—it may be named in the smallest things. All things return (to their root and disappear), and do not know that it is it which presides over their doing so;—it may be named in the greatest things. Hence the sage is able (in the same way) to accomplish his great achievements. It is through his not making himself great that he can accomplish them.[58]

On May 6, 2001, in a formal dialogue with the author on Daoism and science, Professor Edward O. Wilson expressed his deep respect for Lao

Zi's great idea and held that we should explore the legacy of Daoism to find the best path for the future development of science. That important and practical task has been the author's primary focus for the past ten years. Yes, Daoism is partly a religion with an emphasis on immortality. But its deep history is one of exploration of nature and man himself; it aims to transcend by both science and spirituality.[59]

Daoism, the "king of hundred rivers and seas,"[60] is a philosophy generated from and for human nature. It has unique strength and a flexible structure whereby people learn to live a freer life according to their hearts. It will be a helpful guide as we pursue the convergence of science and spirituality with an open mind, working for the ultimate welfare of humankind.

## Notes

1. The key word "Dao" is also translated as "Tao" or "Way" in different texts, but they refer to the same Chinese word "Dao" (in pinyin).
2. Blaise Pascal, *Pensées*, trans. Roger Ariew (Indianapolis: Hackett Publishing, 2005), 64.
3. Ibid.
4. Par Pierre Simon Laplace, *Exposition du Systéme du Monde* (Paris: de l'Institut National de France, et du Bureau des Longitudes, Imprimerie du Cercle Social, l'an IV, 1796).
5. P. Laplace, "Essai Philosophique sur les Probabilités forming the introduction to his *Théorie Analytique des Probabilités* (Paris: V Courcier, 1820); repr., F. W. Truscott and F. L. Emory, trans., *A Philosophical Essay on Probabilities* (New York: Dover, 1951).
6. Lao Zi, *Dao De Jing*, chap. 4.
7. Ibid., chap. 32.
8. Werner Heisenberg, "Über den anschaulichen Inhalt der quantentheoretischen Kinematik und Mechanik," *Zeitschrift für Physik* 43 (1927): 172–98. Translated into English by J. A. Wheeler and W. H. Zurek, published under the title "The Physical Content of Quantum Kinematics and Mechanics," in J. A. Wheeler and W. H. Zurek, eds., *Quantum Theory and Measurement* (Princeton, NJ: Princeton University Press, 1983), 64.
9. Stephen W. Hawking, *The Illustrated Brief History of Time*, updated and expanded ed. (New York: Bantam, 1996), 73.
10. *Dao De Jing*, in *Secret Book of the East*, trans. J. Legge, vol. 39 (Oxford: Clarendon, 1891), chap. 1.
11. Burton Watson, trans., *The Complete Works of Chuang Tzu* (*Zhuang Zi*) (New York: Columbia University Press, 1968), 235.

12. Ibid., 237.
13. Ibid., 240.
14. Albert Einstein, *Relativity. The Special and the General Theory*, trans. Robert W. Lawson (New York: Three Rivers Press, 1961), Appendix Five, "Relativity and the Problem of Space," 155–78.
15. Michael Talbot, *The Holographic Universe* (New York: Harper Perennial, 1992), 51–52.
16. Lee Smolin, *The Life of the Cosmos* (New York: Oxford University Press, 1999), 252.
17. *Dao De Jing*, chap. 41.
18. *Dao De Jing*, chap. 25.
19. Werner Heisenberg, *Physics and Philosophy: the Revolution in Modern Science* (New York: Harper and Row Publishers, 1962), 58.
20. Peng Wenxi, *Xiaoma guo he* [Little Horse Crossing River] (Jiangsu: China Juvenile and Children's Books Publishing House, 1957). A children's picture book. Condensed citation translated by the author.
21. Watson, *Complete Works of Chuang Tzu*, 243.
22. Ernst Cassirer, *An Essay on Man*: *An Introduction to a Philosophy of Human Culture* (New Haven, CT: Yale University Press, 1944), 3.
23. Jose Ortega y Gasset, *History as a System* (New York: W. W. Norton, 1941), 204–5.
24. Ibid., 216.
25. Ibid., 217.
26. David F. Peat, *From Certainty to Uncertainty: The Story of Science and Ideas in the Twentieth Century* (Washington, DC: Joseph Henry Press, 2002), 26.
27. *Dao De Jing*, chap. 2.
28. *Dao De Jing*, chap. 11.
29. Watson, *Complete Works of Chuang Tzu*, 240–41.
30. *Dao De Jing*, chap. 14.
31. Ibid., chap. 21.
32. Ibid., chap. 42. Translated by the author.
33. *Dao De Jing*, chap. 2.
34. Ibid., chap. 77.
35. *Dao De Jing*, chap. 10.
36. Notice that in Lao Zi's statement of "all things have their female and male contents in themselves; in-between them vigorous vitality emerges," the statement of "emergence of vigorous vitality" should come of the observation of heated steaming: fire and water marry into steam, which combines the attributes both of water—wetness and of fire—hotness. Here "steam" is "Three" in *Dao De Jing*.
37. *Dao De Jing*, chap. 40. J. Legge's rhymed translation is: "The movement of the Tao By contraries proceeds; And weakness marks the course Of Tao's mighty deeds," *Secret Book of the East*. Chap. 40.
38. J. B. Hartle and S. W. Hawking, "Wave function of the Universe," *Phys. Rev.* D 28, nos. 12–15 (1983): 2960–75.

39. *Dao De Jing*, chap. 6.
40. Wang Xuanlan, "On the Noumenon of the Dao," in *The Daoist Canon*, vol. 22 (Beijing: Cultural Relics Press, 1988), 886. Translated from ancient Chinese by the author.
41. *Dao De Jing*, in *Secret Book of the East*, chap. 14.
42. Ibid., chap. 21.
43. *Dao De Jing*, chaps. 47 and 48.
44. Watson, *Complete Works of Chuang Tzu*, 118–20.
45. David Bohm, *Wholeness and the Implicate Order* (London: Routledge & Kegan Paul, 1980), 175.
46. Talbot, *Holographic Universe*, 50.
47. Renee Weber, "The Good, The True, The Beautiful: Are They Attributes of the Universe?" *The American Theosophist* 65 (January 1977): 4–13.
48. Watson, *Complete Works of Chuang Tzu*, 293.
49. Ibid., 97.
50. Ibid., 43.
51. Ibid.
52. For related studies, see Norman Girardot, *Myth and Meaning in Early Taoism* (Berkeley: University of California Press, 1983), 150–54; Jiang Sheng and Tong Waihop, eds., *The History of Science and Technology in Taoism*, vol. 1 (Beijing: Science Press, 2002), 659–65.
53. Watson, *Complete Works of Chuang Tzu*, 77.
54. Watson, *Complete Works of Chuang Tzu*, 152–53.
55. Paul Hawken, Amory Lovins, and L. Hunter Lovins, *Natural Capitalism: Creating the Next Industrial Revolution* (Newport Beach, CA: Back Bay Books, 2008), 6.
56. Heisenberg, Physics and Philosophy, 187.
57. Edward O. Wilson, *Consilience: The Unity of Knowledge* (New York: Alfred A. Knopf, 1998), 137.
58. *Dao De Jing*, chap. 34.
59. As early as in 1935, in his book *My Country and My People* (New York: Halcyon House, 1935), Yutang Lin says that Daoism is an attempt for the Chinese to find the secret of nature.
60. Lao Zi says in *Dao De Jing*, chap. 66:

    That whereby the rivers and seas are able to receive the homage and tribute of all the valley streams, is their skill in being lower than they;—it is thus that they are the kings of them all. So it is that the sage (ruler), wishing to be above men, puts himself by his words below them, and, wishing to be before them, places his person behind them. In this way though he has his place above them, men do not feel his weight, nor though he has his place before them, do they feel it an injury to them. Therefore all in the world delight to exalt him and do not weary of him. Because he does not strive, no one finds it possible to strive with him.

## Bibliography

Bohm, David. *Bohm-Biederman Correspondence*. London: Rutledge, 1999.

———. *On Creativity*. London: Routledge, 1998.

———. *Science, Order, and Creativity: A Dramatic New Look at the Creative Roots of Science and Life*. New York: Bantam Books, 1987.

———. *Wholeness and the Implicate Order*. London: Routledge & Kegan Paul, 1980.

Cassirer, Ernst. *An Essay on Man: An Introduction to a Philosophy of Human Culture*. New Haven, CT: Yale University Press, 1944.

*Dao De Jing*. In *Secret Book of the East*. Translated by J. Legge. Vol. 39. Oxford, Clarendon: 1891.

Einstein, Albert. "Inaugural Lecture," July 2, 1914 "Antrittsrede," *Königlich Preußische Akademie der Wissenschaften* (Berlin) *Sitzungsberichte* (1914). In *The Collected Papers of Albert Einstein, Volume 6: The Berlin Years: Writings, 1914–1917*, translated by Alfred Engel. Engelbert Schucking, consultant. Princeton, NJ: Princeton University Press, 1997, 16–19.

———. *Relativity: The Special and the General Theory*. Translated by Robert W. Lawson. New York: Three Rivers Press, 1961.

Fox, Matthew. *One River, Many Wells: Wisdom Springing from World Faiths*. New York: Jeremy P. Tarcher/Penguin, 2004.

Girardot, Norman. *Myth and Meaning in Early Taoism*. Berkeley: University of California Press, 1983.

Hartle, J. B., and S. W. Hawking. "Wave Function of the Universe." *Phys. Rev.* D 28, nos. 12–15 (1983): 2960–75.

Hawken, Paul, Amory Lovins, and L. Hunter Lovins. *Natural Capitalism: Creating the Next Industrial Revolution*. Newport Beach, CA: Back Bay Books, 2008.

Hawking, Stephen W. *The Illustrated Brief History of Time*. Updated and expanded ed. New York: Bantam, 1996.

Heisenberg, Werner. *Physics and Philosophy: The Revolution in Modern Science*. New York: Harper and Row Publishers, 1962.

———. "Über den anschaulichen Inhalt der quantentheoretischen Kinematik und Mechanik." *Zeitschrift für Physik* 43 (1927): 172–198. Translated by J. A. Wheeler and W. H. Zurek, 1981, published under the title: "The Physical Content of Quantum Kinematics and Mechanics." In *Quantum Theory and Measurement*, edited by J. A. Wheeler and W. H. Zurek. Princeton, NJ: Princeton University Press, 1983.

Lao Zi. *Dao De Jing: Secret Book of the East*. Vol. 39. Translated by J. Legge. Oxford: Clarendon, 1891.

———. *Tao Te Ching*. Translated by A. S. Kline. http://www.tonykline.co.uk/PITBR/Chinese/TaoTeChing.htm, 2003.

Laplace, P. "Essai Philosophique sur les Probabilités." Introduction in *Théorie Analytique des Probabilités*. Paris: V Courcier, 1820; repr., F. W. Truscott, and F. L. Emory, trans. *A Philosophical Essay on Probabilities*. New York: Dover, 1951.

Laplace, Par Pierre Simon. *Exposition du Systéme du Monde*. Paris: de l'Institut National de France, et du Bureau des Longitudes, Imprimerie du Cercle Social, l'an IV, 1796.

Lin, Yutang. *My Country and My People*. New York: Halcyon House, 1935.

Ortega y Gasset, Jose. *History as a System*. New York: W. W. Norton, 1941.

Pascal, Blaise. *Pensées*. Translated by Roger Ariew. Indianapolis: Hackett Publishing, 2005.

Peat, David F. *From Certainty to Uncertainty: The Story of Science and Ideas in the Twentieth Century*. Washington, DC: Joseph Henry Press, 2002.

Peng, Wenxi. *Xiaoma guo he* [Little Horse Crossing River]. Jiangsu: China Juvenile and Children's Books Publishing House, 1957.

Sheng, Jiang, and Tong Waihop, eds. *The History of Science and Technology in Taoism*. Vol. 1. Beijing: Science Press, 2002.

Smolin, Lee. *The Life of the Cosmos*. New York: Oxford University Press, 1999.

Talbot, Michael. *The Holographic Universe*. New York: Harper Perennial, 1992.

Wang, Xuanlan. "On the Noumenon of the Dao," in *The Daoist Canon*. Vol. 22. Beijing: Cultural Relics Press, 1988.

Watson, Burton, trans. *The Complete Works of Chuang Tzu*. New York: Columbia University Press, 1968.

Weber, Renee. "The Good, The True, The Beautiful: Are They Attributes of the Universe?" *The American Theosophist* (Wheaton, Illinois) 65 (January 1977): 4–13.

Wilson, Edward O. *Consilience: The Unity of Knowledge*. New York: Alfred A. Knopf, 1998.

Editor's Introduction to

# Whitehead Reconsidered from a Buddhist Perspective

**Ryusei Takeda,** Ryukoku University, Japan

Ryusei Takeda challenges an important strand of philosophical theology that is widely applied in analyses of our relationship with the physical universe—"Process Thought." At once taking exception to Alfred North Whitehead's introductory critique of Buddhism and crediting him with building an important complementary idea system, Takeda offers a novel and compelling vision of the synergy between Buddhist thought and process theology.

In directly referring to Buddhism, Whitehead aligns it with his typology's third strain of thought about the image and nature of God. He holds that it emerges "in the image of an absolute philosophical principle." But Takeda argues that, in fact, the essential compassion and soteriology of Mahayana Buddhism is more reminiscent of Whitehead's fourth "Galilean" strain, as present in the origin of Christianity. Buddhism demands, as does Whitehead, that temporality be reinterpreted "based in the ultimate principle that derives from both the eminently real and the unmoved mover." Takeda asks, "What type of creative transformation must occur through the positive, creative meeting of Buddhism and Christianity"?

If Christianity is a religion seeking a metaphysics and Buddhism a metaphysic generating a religion, as Whitehead claims, it must also be noted that Gautama Buddha was intentionally silent on key metaphysical points. This silence, and the careful path of attention to "suchness" embedded in Buddhist practice is not the adherence to supposedly self-evident metaphysical ideas that Whitehead criticized.

For although overattention to the historical and doctrinal might imperil a tradition with stagnation, Mahayana Buddhism is vital, with its foundation in "the ultimate place of self-existence, realized as the primordial fact of the inquiry into and clarification of the self."

Whitehead ignores this foundational mode and criticizes Buddhism as static and unfruitful. He entirely misses the coherent effort toward the "realization of the wisdom of *prajna*, that is the attainment of discrimination after the negation of non-discrimination," the capacity to critically analyze after passing through a state of acritical-ness. But Whitehead's concern is not without merit and similar critiques appear directly within the Mahayana tradition, coupled with a caution of the vehicles toward enlightenment—the disciple and the lone seeker—becoming bound up in these levels is a trap of inescapable stagnation.

Takeda demonstrates that Mahayana Buddhism operates on a dynamic basis that embraces the "interrelatedness of all things." In the context of compassion and with care to avoid static-ness, Buddhism offers a vision similar to Whitehead's of the mutuality of coming into being.

P.D.

6

# Whitehead Reconsidered from a Buddhist Perspective

**Ryusei Takeda,** Ryukoku University, Japan

Both fundamental differences and important resemblances can be found between process metaphysics and Buddhist thought. In considering Alfred North Whitehead, perhaps the foremost modern representative of process metaphysics, we must ask, from the standpoint of his own interpretations of process philosophy, how he grasps "Buddhist thought"—that is, does he understand it or criticize it? (For those who are interested in my observations of another major process thinker, Charles Hartshorne, please see my Japanese article in *Process Thought*, the journal for the Japan Society for Process Studies.)

I will also consider from a Buddhist point of view how Whitehead's interpretation of Buddhism can be evaluated in terms of its significance for Buddhism. There is a rich development of a deep suggestion for engaging modern scientific knowledge underlying Whitehead's view of Buddhism, and I want to illuminate and draw out one side of this task that modern Buddhist thought must acknowledge.

Whitehead refers to Buddhism three times in *Process and Reality*. The first time, in part III, Buddhism appears together with Greek thought. Namely, Whitehead postulates a basic enjoyment, "the initial stage of its aim [being] an endowment which the subject inherits from the inevitable ordering of things, conceptually realized in the nature of God" (*PR*, 244). Whitehead insists that "the remorseless working of things in Greek and in Buddhist thought" resembles the "inevitable ordering of things, conceptually realized in the nature of God" (*PR*, 244). This points to such ideas found in Buddhist thought as transmigrating through the cycle of birth-and-death (*samsara*) and the chain of causation through karma, whether evil or good.

Further, the second and third references to Buddhism occur together in part V, where Whitehead again takes up the temporal world:

> So long as the temporal world is conceived as a self-sufficient completion of the creative act, explicable by its derivation from an ultimate principle which is at once eminently real and the unmoved mover, from this conclusion there is no escape: the best that we can say of the turmoil is, "For so he giveth his beloved sleep." (*PR*, 342)

For Whitehead, Buddhism does not go beyond this.

Within this matter, although he recognizes that "in some sense it is true," he also insists that "we have to ask, whether metaphysical principles impose the belief that it is the whole truth." Here Whitehead is emphasizing the need to "fathom the deeper depths of the many-sidedness of things" that cause one to consider the complexity of the world. We can envision Whitehead critiquing the dogmas of Buddhism as simply seeking the world of the peace of mind of nirvana. In Whitehead's criticism, Buddhism is seen as childish; however, can we not instead regard this criticism itself as simplistic?[1]

Still further, Whitehead considers the "three strains of thought" related to God's image to each emerge from the "great formative period of theistic philosophy" (*PR*, 342). The third such strain he describes as the concept of God that formed "in the image of an absolute philosophical principle," which he associates with Aristotle, noting however that "Aristotle was antedated by Indian, and Buddhistic, thought" (*PR*, 343). Here as well, the fundamental factor of Buddhist thought is that it is in some way philosophical, namely, that it is comparable to Aristotelian thought, and can thus be thought of as a principle positioned at the highest level.

In taking the "ultimate philosophical principle" as distinct from the other two strains of thought—namely the "image of the imperial ruler" associated with Caesar and the "image of a personification of moral energy" associated with the Hebrew prophets—and being instead like the cold fact of reason, which contains neither the element of the ruler nor the primary factor of ethical characterization, we can say that Whitehead holds the fundamental position of Buddhism. And further, there is what can be called a fourth strain, pointing to the "Galilean origin of Christianity," which "does not fit very well" with these three strains of thought, and which is characterized by Whitehead as follows: the suggested image of God with its Galilean origin

> dwells upon the tender elements in the world, which slowly and in quietness operate by love; and it finds purpose in the present immediacy of a kingdom not of this world. Love neither rules, nor is it unmoved; also it is a little oblivious to morals. It does not look to the future; for it finds its own reward in the immediate present. (*PR*, 343)

This image of God seen in this Galilean origin of Christianity is surprisingly close to the image of the other-benefiting great compassion found in the bodhisattva path of Mahayana Buddhism. This image of the compassion of the bodhisattva path is seen especially within the image of the Buddha-body extending to both the causal and resultant stages of Amida Buddha, the Tathagata who has appeared, in his boundless great compassion, filled with commiseration for the beings in the world, and expounded the teachings of the way to enlightenment, seeking to save the multitudes of living beings by blessing them with the benefit that is true and real. We can say that the concern with a world of deep love found in the Galilean origin of Christianity, namely, the two dynamic factors of the "tender elements in the world" and the "present immediacy of a kingdom not of this world," together occupy an essential, foundational position in this image of Buddhist compassion as well.

So far we have examined Whitehead's view of Buddhism in process and reality. We can sum it up in the following two points:

The first point concerns the impermanent world of temporal transmigration. Speaking from a Buddhist point of view, how should its significance be understood? According to Whitehead's critique, the temporal world, explained by and based in the ultimate principle that derives from both the eminently real and the unmoved mover, must be reinterpreted from within various metaphysical principles.

The second point is a problem connected to the relationship with Christianity. According to Whitehead's criticism, Buddhist thought is simply referred to as an ultimate philosophical principle, but can the basic perspective of Buddhism really be understood as such? We must undertake the important task of considering what type of creative transformation must occur through the positive, creative meeting of Buddhism and Christianity.

These two aspects of Whitehead's view of Buddhism have been developed further in the original views of Buddhism of two people: the first view regarding metaphysics has been developed by Charles Hartshorne, while the second view regarding the relationship between Buddhism and Christianity has been clarified by Dr. John B. Cobb.

The references to Buddhism in Whitehead's *Religion in the Making* consist mainly of comparisons to Christianity and discussions of the continuing significance of the two religions. First, in relation to evil, Buddhism

> finds evil essential in the very nature of the world of physical and emotional experience. The wisdom which it inculcates is, therefore, so to conduct life as to gain a release from the individual personality which is the vehicle for such experience. The Gospel which it preaches is the method by which this release can be obtained. (*RM*, 49)

However, "one metaphysical fact about the nature of things which it presupposes is that this release is not to be obtained by mere physical death. Buddhism is the most colossal example in history of applied metaphysics" (*RM*, 49–50).

Whitehead's discussion of comparative religion, in which he sees Christianity as "a religion seeking a metaphysic, in contrast to Buddhism which is a metaphysic generating a religion" (*RM*, 50), concisely expresses one of the most essential, fundamental differences between the two religions. The fact is, we can also say that in the case of Mahayana Buddhism, and particularly the Pure Land teachings, Lotus Sutra thought, and Shingon esotericism that developed out of Gautama Buddha's doctrines, religious form was born from metaphysics. However, from the viewpoint that Gautama Buddha kept silent in opposition to the metaphysical questions regarding such issues as whether or not the universe is eternal, finite, or infinite, or what happens to a Buddha after death (*avyakrta*), the often-indicated viewpoint of the negation of metaphysical principles does not match Whitehead's understanding.

In my view the Buddha never negated metaphysics. The Buddha started by looking directly at the reality of the suffering of human existence and awakened through deep meditation to the fundamental cause of that suffering as being rooted in the ignorance of human existence. Thereafter he preached the path to emancipation (*vimoksha-marga*), vowing that all sentient beings awaken to that wisdom of emancipation (*vimukti-jnana*). Here we find the origin of the three modes of learning, that is, morality, meditation, and wisdom (*shila*, *samadhi*, and *prajna*), the fundamental systematization of the Buddhist path to nirvana.

The basic spirit that constantly penetrates the Buddha's path to enlightenment is none other than the awakening to true suchness (*evam-jnana, evam-darshana*), the awakening to the true thusness of reality. I feel that this is also the basic stance of Whitehead, the philosopher—what we can call the metaphysics of his so-called "Philosophy of Organism." I dare to call this the "philosophy of *pratityasamtupada*."

Whitehead's strict criticism, the point that metaphysics can easily fall into dogma (*PR*, 9), could be the greatest reason for the Buddha's silence in answer to metaphysical questions. For Whitehead, metaphysics is "the endeavour to frame a coherent, logical, necessary system of general ideas in terms of which every element of our experience can be interpreted" (*PR*, 3).[2]

Actually, according to his regulations of metaphysics, Whitehead asserts that a general defect of metaphysical systems "is the very fact that it is a neat little system of thought, which thereby over-simplifies its expression of the world" (*RM*, 50). Here Whitehead finds a reason that Buddhism was not able to develop to the extent of Christianity. Whitehead holds the idea that Gautama Buddha's metaphysical ideas are announced as self-evident, and doctrinal development, in the end, is based in none other than the original doctrine.

This viewpoint of Whitehead's is correct in a sense. However, from the other side Whitehead overlooks important core parts of Buddhism. Buddhism is, to the end, a religion of practice. Through practice, one reaches a self-transformational awakening, the aim being a subjective awakening to the true thusness of the reality of all things. The significance is that one cannot dogmatically cling to Gautama Buddha's teachings. Further, Whitehead states that "Buddhism starts with the elucidatory dogmas; Christianity starts with the elucidatory facts" (*RM*, 52). Namely, he presumes that "the Buddha left a tremendous doctrine. The historical facts about him are subsidiary to the doctrine" (*RM*, 51). In contrast to this, Christianity "starts with a tremendous notion about the world. But this notion is not derived from a metaphysical doctrine, but from our comprehension of the sayings and actions of certain supreme lives. It is the genius of the religion to point at the facts and ask for their systematic interpretation" (*RM*, 50). Here, the fundamental difference between both religions, that is, metaphysical doctrine and historical religious facts, can be seen in the categories concentrated in the opposition between theory and life. This view of Buddhism held by Whitehead, seen from the standpoint of the Mahayanists who rose about four hundred years after the Buddha passed into nirvana, also cannot be accepted.

From a Mahayana point of view, Whitehead emphasizes the metaphysical system and doesn't recognize the development of Buddhism. The historical truth of the Buddha, that is, his becoming an awakened one accomplishing the vows and practices of the bodhisattva during the causal stage and attaining *mahaparinirvana*, is the most significant point, and the various doctrines that were developed up until that time (that is, *abhidharma*, etc.) turned out to be auxiliary. Mahayana Buddhists reinterpreted the historical fact of the Buddha in such a way that they realized that Gautama Buddha's enlightenment is immanent within the intrinsically pure self-nature of all sentient beings. This reinterpretation was conceptualized as the Mahayana Buddhist scriptures.

Therefore, concerning the problem of how both religions prove the emancipation from evil, Whitehead sees Buddhism analyzing the reality of evil and establishing its doctrine, after which it provides certainty. Rather, it is the case that the Mahayana Buddhist conquest of evil is sought by awakening to the formless life potential of taking any form immanent in the Buddha nature within the self. This is not a metaphysical theory. It is the fundamental fact of the solemn awakening to the Buddha nature. That is, the bottomless foundation of the ultimate place of self-existence, realized as the primordial fact of the "inquiry into and clarification of the self." The Buddhist conquest of evil is none other than this.

Such a Mahayana Buddhist principle of awakening to the Buddha nature rooted in the real historical existence of the self is spoken of in the same way by Whitehead in terms of what he calls the "final principle of religion": "The final principle of religion is that there is a wisdom in the nature of things, from which flow our direction of practice, and our possibility of the theoretical analysis of fact" (*RM*, 137–38). Further, Whitehead states:

> Religions commit suicide when they find their inspirations in their dogmas. The inspiration of religion lies in the history of religion. . . . The sources of religious belief are always growing. . . . Records of these sources are not formulae. They elicit in us intuitive response which pierces beyond dogma. (*RM*, 138–39)

Here we see that Whitehead's theory of religion does not differ fundamentally from the basic standpoint of Mahayana Buddhism.

Having above stated my critique of Whitehead's view of Buddhism, Whitehead's next words are suggestively prophetic. Here we catch a glimpse of Whitehead's insightful view of Buddhism:

> Buddhism and Christianity find their origins respectively in two inspired moments of history: the life of the Buddha, and the life of Christ. The Buddha gave his doctrine to enlighten the world: Christ gave his life. It is for Christians to discern the doctrine.
>
> Perhaps in the end the most valuable part of the doctrine of the Buddha is its interpretation of his life. (*RM*, 55)

It can be said that the historical significance at the core of Mahayana Buddhism is just such a subjective interpretation of the life of the Buddha.

The final issue treated in Whitehead's criticism of Buddhism and Christianity in *Religion in the Making* points to the cause of the decline of both religions. According to Whitehead, this decline is due in part to the closed nature of both religions based in their self-sufficient philosophies, as well as the fact that "both have suffered from the rise of the third tradition, which is science, because neither of them had retained the requisite flexibility of adaptation" (*RM*, 140–41). Although he severely criticizes both religions on this point, he does not simply reject them, but instead makes the positive suggestion that we can find a deeper meaning within each. Both Buddhism and Christianity must deal with this theme on their own. In recent years international conferences regarding the issue of dialogue between the world religions have gradually become more and more active. However, in the Buddhist world one cannot deny that dialogue has a long way to go. John B. Cobb's view of Buddhism announces positive possibilities for the dialogue between Buddhism and Christianity, while Charles Hartshorne's view of Buddhism offers important suggestions for the Buddhist side.

Next, I would like to consider Whitehead's view of Buddhism recorded in a series of dialogues by Lucien Price. Here, Whitehead gives the following critical view of Buddhism in his explanations of religion: it is static (*D*, 164), "a religion of escapism" (189), in which one becomes absorbed in "an unfruitful passive mediation" (301), and it teaches that "we must keep returning lifetime after lifetime for purification through experience until we are worthy to lose our identity in the all" (198). Further, Whitehead declares that, as a Buddhist, "you retire into yourself and let externals go as they will. There is no determined resistance to evil. Buddhism is not associated with an advancing civilization" (189). Whitehead himself is not in any way satisfied with Buddhism, in the end criticizing it as a religion that causes social stagnation (301).

These dialogues of Whitehead were recorded from 1941 through 1944, a time when Buddhist studies in the West were not yet developed sufficiently. As a Buddhist, it would be easy to reject these criticisms of Buddhism, as they expose his ignorance of the historical development of Mahayana Buddhist thought. In actuality, Buddhist thought succeeded to ancient Indian tradition, and even while the intentions and aims may differ, the practice of meditation (*dhyana*, *samadhi*) is the essence of the path of practice toward awakening and emancipation in Mahayana Buddhism. It is here that one awakens to the realization of the wisdom of *prajna*, that is, the attainment of discrimination after the negation of nondiscrimination. It is not at all the case that Buddhists wish to escape from the world or bring about social stagnation. As I stated before, the intention of the Mahayana Buddhist path is the exact opposite of Whitehead's view of Buddhism. However, Buddhism must respond to the various social, political, economic, and philosophical problems of the modern world more quickly than it has in the past. The immanent transcendence that faces toward absolute emptiness and drops off the dichotomy between body and mind in the bottomless foundation of one's own existence is not the cause of the simple noninvolvement of the retreat into oneself and lack of concern with the outside world pointed out by Whitehead, nor does Mahayana Buddhism aim toward a life of seclusion. Rather, the awakening to the absolute emptiness at the subjective source of the self is accompanied by a thorough involvement with all worldly matters. If the term "*soku*" in the expressions "*samsara* is nirvana" and "the passions are enlightenment" does not indicate a state of silence, then neither does it lead to social stagnation; instead, it indicates proper engagement with the world of the passions and birth and death.

Actually, a sharp critique similar to Whitehead's is found in Mahayana Buddhism as well. The dangerous possibility of the bodhisattva falling back to the level of the two vehicles of the *shravaka* and the *pratyekabuddha* is pointed out in the teachings. For the bodhisattva, dropping to the level of the two vehicles is more frightening than falling into hell. Even if one falls into hell, the attainment of Buddhahood is still possible, but since dropping to the level of the two vehicles results in the impossibility of attaining Buddhahood, it came to be referred to as the "death of the bodhisattva." Therefore, the criticism of Buddhism that occurs in Whitehead's writings could be said to be a criticism of the two vehicles of the *shravaka* and the *pratyekabuddha*. At the same time, the view of religion found in Whitehead's own philosophy of organism within which his criticism is developed is, if I dare say so, a clarification of the structure of the philosophical analysis of the religious world of the identity and negation of *prajna* arrived at by the Mahayana bodhisattva path.

In Whitehead's discussion of the finite and the infinite in *Essays in Science and Philosophy*, he states that Buddhism "emphasizes the sheer infinity of the divine principle, and thereby its practical influence has been robbed of energetic activity" (*ESP*, 106). According to Whitehead, Spinoza "emphasized the fundamental infinitude and introduced a subordinate differentiation by finite modes" (106). In opposition to this, Leibniz "emphasized the necessity of finite monads and based them upon a substratum of Deistic infinitude" (106). However, viewed from Whitehead's perspective, neither one of them understood the precise relationship between the infinite and the finite. In Whitehead's statement that "infinitude is mere vacancy apart from its embodiment of finite values, and . . . finite entities are meaningless apart from their relationship beyond themselves" (106) can be seen the simultaneous nonidentity and nonseparation between finitude and infinity.[3]

From the above discussion we can see that Whitehead's criticism of Buddhism is not adequate in the case of Mahayana Buddhism. Rather, Whitehead's own standpoint can be said to approximate the Mahayana relationship between the finite and the infinite. For example, this is adequately indicated in the relationship between identity and negation in the expressions "the passions are enlightenment" and "*samsara* is nirvana." However, as Whitehead's criticism indicates, among Buddhists there is an undeniable tendency to see the ultimate reality within the infinite rather than the finite.

I hope that I have shown not only where Whitehead's understanding of Buddhism falls short, but more importantly, his perceptive criticisms, which Buddhists must take to heart. Although Mahayana Buddhism stresses emptiness, its notions of ultimate reality, if taken as substantial, become static. And even though Whitehead's view of reality as process does not explicitly contain any Buddhist notions of suchness or emptiness (*shunyata*), it can still provide Buddhists with some important insights. Nagarjuna's intent in introducing *shunyata* was to negate clinging, which is caused by our normal way of looking at reality as substantial or unchanging. If one takes reality as process, then automatically such clinging is negated. Though certainly not identical, Whitehead's philosophy can potentially clarify the interrelatedness of all things. But this needs to be explored more fully, based on *shunyata* and *pratityasamtupada*, without returning to the substantialist understanding of *abhidharma*. Ultimately, no human activity can clarify this relationship. But the sutras attempt to express this true nature of reality in language, even though we are aware of the limitations. Whitehead has given us another possible means of understanding and explicating these fundamental Buddhist ideas.

## Notes

1. As a work that gathers the understandings of nirvana by Westerners, Guy Welbon's book, *The Buddhist Nirvana and Its Western Interpreters* (Chicago: University of Chicago Press, 1968) is well known. However, while using this work as a base, John B. Cobb, in his *Beyond Dialogue: Toward a Mutual Transformation of Christianity and Buddhism* (Minneapolis: Fortress Press, 1982), 55–74, attempts his own keen philosophical interpretation of nirvana while maintaining a critical consideration of Western scholars' understandings.
2. Beyond this, metaphysics has the realistic intent of not only analyzing so-called metaphysical propositions, but the strict analysis of various propositions of use in everyday life. Further, that main role includes the clarification of the meaning of the expression "all things flow" (*PR*, 208). Also, regarding the relationship with religion, he states,

   That we fail to find in experience any elements intrinsically incapable of exhibition as examples of general theory is the hope of rationalism. This hope is not a metaphysical premise. It is the faith which forms the motive for the pursuit of all sciences alike, including metaphysics. (*PR*, 42)

   It is here that we find "the point where metaphysics and indeed every science gains assurance from religion and passes over into religion" (*PR*, 42). Finally, we must listen closely to Hartshorne's insistence that "probably the most important function of metaphysics is to help in whatever way it can to enlighten and encourage man in his agonizing political and religious predicaments" (Charles Hartshorne, *Creative Synthesis and Philosophic Method* [1970; repr., Chicago: Open Court Publishing Co., 1983], 55).
3. Regarding this connection, Nishida Kitaro defines "the infinite" as follows:

   The infinite (*das Unendliche*) is not simply the negation of the finite, it is not simply *das Endlose*. Thus, the infinite is [actually] Hegel's so-called "*schlechte oder negative Unendlichkeit*," and can instead be called finite. The truly infinite stores the motive for transformation in the self, it is the development of differentiation in the self. (Shisaku to taiken, "Riron no rikai to suri no rikai," in *Nishida Kitaro Zenshu*, vol. 1. 4th ed. (Tokyo: Iwanami, 1987–89), 263–64.

Further, regarding the formation of the infinite series, Nishida states:

> That which is called knowing must first of all be a containment inside. However, when that which is contained is conceived of as the container, in the same way that we can think of a physical object in space, it is none other than mere spatial existence. When we can think of the container and that which is contained as the same thing, it is formed in the same way as the infinite series. Also, when we can think of that single thing containing infinite mass within the self, then we can think of that which is working infinitely as pure function. (Hatarakumono kara mirumono e, "Basho," in *Nishida Kitaro Zenshu*, vol. 4, 215–16)

## Bibliography


Price, Lucien. *Dialogues of Alfred North Whitehead*. Westport, CT: Greenwood Press, 1977 (originally published in 1954). Abbreviated as *D*.

Whitehead, Alfred North. *Essays in Science and Philosophy*. Westport, CT: Greenwood Press, 1968 (originally published in 1947). Abbreviated as *ESP*.

———. *Process and Reality*. Corrected ed. Edited by David Ray Griffin and Donald W. Sherburne. Washington, DC: Free Press, 1978 (originally published in 1929). Abbreviated as *PR*.

———. *Religion in the Making*. New York: New American Library, 1974 (originally published in 1926). Abbreviated as *RM*.

## Editor's Introduction to

# Sanctity of Life: A Reflection on Human Embryonic Stem Cell Debates from an East Asian Perspective

**Heup Young Kim,** Kangnam University, South Korea

As genetic technology races forward, we find that important intercultural differences are emerging in areas of human rights, differences that have remained largely unexplored. In human cloning, for example, Asian spiritual traditions present substantially different contexts than the West's Abrahamic religions. Heup Young Kim frames the issue of human embryonic stem cell research in a hermeneutic that spans Christian theology and ethics as well as Confucian and Taoist thought. Personhood and respect figure in both dialectics, and he points to a close correlation between the proleptic, teleological origin of the sanctity of human life in Christian and Confucian contexts. Furthermore, both Western and Asian thought reflect the imperative of self-realization and offer pathways toward a humble, anthropocosmic trajectory.

Kim opens his analysis by framing human embryonic cell research as a *koan,* an "evocative question" around which to organize a new hermeneutic for the "biotech century." He begins in a Western context by critiquing the tendency to focus on the "micro" issue of the curative potential of new therapies at the expense of closer scrutiny of the "macro" questions of cultural consequences. He goes on to point out the power of rhetorical control, the self-conscious elite intention to reify issues along particular sociomoral lines. Finding the initial debate in the United States to be a tug-of-war over this process of rhetorical construction, he notes that much of the controversy so far has focused on simple distinctions like the aliveness or mechanism of a blastocyst, distinctions of a type that is conducive to the familiar counterarguments for beneficence or nonmalificence.

The dignity of human embryonic cells marks an essential line in many Western critiques of cell research. Kim traces this definition of dignity—that humans must never be treated as a means to some end—to Kant. And he cites current Catholic doctrine as closing the door completely on human embryonic cell research. But the fundamental problem with the Catholic position is that it concatenates ensoulment with genetic uniqueness, thereby radically questioning the dignity of monozygotic twins, for example, and ignoring the uniformitarian science that shows ontogeny to be a slow, phased process lacking a unique moment of "conception." This Christian position on human dignity begs the questions of innate vs. conferred value. Kim claims that the strongest argument in favor of unique human dignity is, in fact, the future-oriented expectation of resurrection and unity with Christ.

Kim cites Confucian doctrine in arguing that the "great man" rejects a cleavage between self and others, that a "profound ecological sensitivity" is embedded in the guiding virtue of *gyeong*. So caring for all life is a reflexive element of righteous being and does not require an extraneous infusion of "dignity." Likewise, the teleological construction of individual value is modified by the essential Confucian precept that replaces "self-fulfillment" with "being-in-relationship," *imago Dei* with *T'ien-ming* (the holy obligation to work for the well-being and moral order of the universe).

P.D.

# 7

# Sanctity of Life

## A Reflection on Human Embryonic Stem Cell Debates from an East Asian Perspective

**Heup Young Kim,** Kangnam University, South Korea

> Human survival requires both the act of defining and the responsible action that flows from definition. That is what it means to be created co-creator. This self-definition, itself both reflective and political in character, configures the encounter with transcendence in our lives.[1]

### Political Hermeneutics of Life: A Contemporary *Koan*[2]

This provocative statement of Philip Hefner has become a fulfilled prophecy in the arena of international politics. On March 8, 2005, the General Assembly of the United Nations (UN) adopted the Declaration on Human Cloning, "by which Member States were called on to adopt all measures necessary to prohibit all forms of human cloning inasmuch as they are incompatible with human dignity and the protection of human life."[3] This controversial declaration brings about a new battle of definition pertaining to life; in other words, it places the political hermeneutics of life at the center of international geopolitics. After voting against it, the representative from Korea stated: "The term 'human life' meant different things in different countries, cultures and religions. [Thus,] it was inevitable that the meaning of that ambiguous term was subject to interpretation."[4] The representatives from China, Japan, and Singapore made similar responses. This dialogue is reminiscent of the case when Asian, traditionally Confucian, countries advocated Asian values vis-à-vis the human right issues initiated by the United States. Whereas the latter case is related to social ethics (macro), the former case is related to bioethics (micro).

At any rate, life is no longer only an academic, metaphysical subject matter but a concrete concept related to a geopolitical realism of power. The dignity or sanctity of life has become a great *koan* for this century, a "biotech century." The central issue at this moment is the human embryonic stem (hES) cell research. As in the UN declaration, contemporary hES cell debates include four key words, namely, (the sanctity and protection of human) life, (the compatibility with human) dignity, respect, and (the prevention of the exploitation of) women.

The hES cell debate since the beginning, particularly in the case of the United States, has been heavily rhetorical, for it is directly related to government funding and public policy for human health. It is said that the rhetoric for the research is characterized by exaggeration and inflation of promises, and some called it "gene-hype."[5] In fact, there is "the immense and substantive gap between discovery and cure." This high-tech, time-consuming, labor-intensive, and extremely expensive research is much more relevant to "the wealthier areas of the society, where the money is to be made."[6] It holds the position that disease is an individual problem (of the genetic code), neglecting its social and environmental causes. The question of who will benefit from stem cell research is crucial and is not adequately addressed. "The rhetoric is that all will benefit. . . . [However,] even should stem cell-based therapy prove successful, the number of people who stand to benefit from it are a small subset of the whole population and perhaps even a small subset of all those with genetic diseases."[7] In fact, the macro issue is more important than the narrow micro issue. Thomas Shannon argued:

> The micro ethical debate over the use of early human embryos is not the key factor in resolving the larger stem cell debate. Although a case can be made for the use of such stem cells, another more critical variable is the consequence of

objectification of human nature in this way. In principle an argument for the use of such cells exists, the consequences of such use might be more problematic than we realize. However, I think the more important point is the macro issue, the social context in which such cells would be used.[8]

Furthermore, stem cells have been regarded as a subject of "expert bioethics" that necessitates "professional discourse."[9] "Public debate has been minimal," and the rhetoric has been dominated by elites and experts in the biomedical industrial complex. In this politics of rhetoric, "whoever captured the definition of hES cell research had won half the battle."[10] Also, scientific expertise has been used "as a weapon to control definitions."[11] Moreover, differentiating hES cell research from human cloning is a rhetorical strategy to avoid the emotionally charged cloning issue. The experts seem to have learned the following lesson from the cloning controversy: "*in modern biotechnological controversies, public debate must be shepherded and fostered by an elite that is prepared to seize rhetorical primacy, and to mold existing institutions or create new ones for that purpose*."[12] Wolpe and McGee attempted to liberate the stem cell discourse from the rhetoric of expert domination. "The process of deciding who will refine, reform, or reify definitions of these cells is a sociomoral exercise that has implications for the broader battle for or against hES cell research."[13]

### Is the Embryo Life?

The initial, blunt debate in the United States is about whether the embryo is a person or a property. On the one hand, most scientists and supporters for hES cell research claim that an embryo cannot be regarded as a person, but rather as a property (or a cell mass), because it has not yet attached to the uterine wall and gastrulation has not yet occurred (the fourteen-day rule). On the other hand, strong oppositions come from Roman Catholic and conservative Protestant churches. The Vatican's position is most resolute and clear-cut. The embryo possesses full human personhood, dignity, and moral status from the moment of fertilization. Therefore, it is not permissible to harm and destroy the blastocyst (trophectoderm) to derive stem cells from the inner cell mass.

Both positions are extreme and problematic. More careful, sophisticated, but still ambiguous discussions have been generated by the National Bioethics Advisory Commission (NBAC) that the U.S. government established upon the request of President Clinton in 1998. The NBAC proposed the following, more moderate position. The embryo is a form of human life, but not a person (a human subject) yet. So it is not a property but needs to be treated with respect (though not of the same level as a person). This position further developed a delicate distinction between totipotency and pluripotency. Simply, embryos are totipotent, while stem cells are pluripotent. A totipotent embryo (a potential human person) cannot be a subject for research, because it requires harming and destruction. But a pluripotent hES cell can be a subject for research, because it is not an embryo, so not a potential person. This rhetoric of stem cell research is ambivalent, because the nonindividuated embryo *ipso facto* cannot be an individual person.

Another issue in the debate involves a distinction between the principles of beneficence and nonmalificence. It resembles the controversy between the utilitarian ethos and the principle of equality of protection on the issue of the abortion of fetuses. On the one hand, the beneficence (or healing opportunity) position argues for hES cell research for the sake of utilitarian benefits for human health and well-being. On the other hand, the nonmalificence (or embryo protection) position opposes hES to protect the dignity of the embryo (do not harm!), the most vulnerable form of human life. It criticizes that the destruction of blastocysts is a devaluation of human life, and is, so to speak, a kind of infanticide or even a new eugenics of euthanasia. This position presupposes that the human embryo is the tiniest human being and so has dignity. This understanding raises the issues of defining and understanding human dignity and personhood.

### What Is Human Dignity?

The definition of dignity frequently used in stem-cell debates in the United States derives from the philosophy of Immanuel Kant. Treat "each human being as an end, not merely as some further end."[14] This position is also the dominant view of Christian churches. The United Church of Canada elaborated, "In non-theological terms it means that every human being is a person of ultimate worth, to be treated always as an end and not as a means to someone else's ends."[15] The conservative Southern Baptist Convention explicitly identifies human embryos as "the most vulnerable members of the human community."[16] Even the liberal United Methodist Church takes a similar position against hES cell research: "Such practices seem to be destructive of human dignity, and speed us further down the path that ignores the sacred dimensions of life and personhood and turns life into a commodity to be manipulated, controlled, patented, and sold."[17] The Vatican's position is obvious: "The ablation of the inner cell mass of the blastocyst, which critically and irremediably damages the human embryo, curtailing its development, is gravely immoral and consequently gravely illicit."[18]

The underlying logic behind the Vatican's argument is "a tacit association" between "ensoulment" (the infusion of the spiritual soul into the physical body) and "genetic uniqueness" (a new genome) established at conception.[19] Following Pius XII, Pope John Paul II affirmed, "If the human body takes its origin from pre-existent living matter [evolution], the spiritual soul is immediately created by God."[20] In the *Evangelium Vitae* (1995), he stipulated:

> The Church has always taught and continues to teach that the result of human procreation, from the first moment of its existence, must be guaranteed that unconditional respect which is morally due to the human being in his or her totality and unity in body and spirit: The human being is to be respected and treated as a person from the moment of conception; and therefore from that same moment his rights as a person must be recognized, among which in the first place is the inviolable right of every innocent human being to life.[21]

Hence, the door for hES cell research in the Roman Catholic Church is completely closed: "Intentional destruction of innocent human life at any stage is inherently evil, and no good consequence can mitigate that evil."[22]

However, the tacit association between ensoulment and genetic uniqueness has been seriously challenged. First of all, embryology denies the actuality of "a moment of conception," but proves that "conception is a process" for near two weeks leading to implantation.[23] Further, "the blastocyst is not an individual person but a potential person." A potential person cannot be regarded as an actual person just as an acorn is not an actual oak tree. The Vatican's view assumes that genetic uniqueness is the basis for human dignity. But this is hardly tenable, because twinning is a natural phenomenon that occurs prior to fourteen days after fertilization. Mistakenly, it denies the human dignity of monozygotic twins by identifying their undeniable existence as a result of genetic abnormality. Furthermore, it has no room to acknowledge the dignity of cloned people who might appear in the future.

The Western notion of human dignity is primarily based on two pillars: the Christian view of the sanctity of the human person and the Enlightenment idea of the intrinsic value of a human person. Christian theology claims a human person as "an everlasting object of God's love," and Kant attributes self-determination or autonomy as the central value of human personhood.[24] However, both positions view dignity basically as intrinsic, still maintaining problems of Western anthropology such as substantialism, individualism, anthropocentricism, and archonic thinking (morality is grounded in a suitable interpretation of origins).

### What Does Human Person Mean?

Ted Peters suggested three models of Christian anthropology: person as innate, person in communion, and person as proleptic.[25] The first, most prevailing, but defective model discussed in the previous section presupposes dignity as intrinsic or innate. Theologically, dignity is not only intrinsic but also conferred, because human dignity is ultimately endowed by God. "Dignity is first conferred, and then claimed."[26] But Western thought since the Enlightenment assumes that human dignity is inherent, so present at birth. In the genetic age, this archonic thinking generates such a view as the Vatican's that human dignity refers to the genetic uniqueness established at the moment of ovum fertilization and zygote creation.

The second model believes that dignity is not just inborn, but rather it, at least in the sense of self-worth, is relational, "the fruit of relationship."[27] The logic of genetic uniqueness "cannot count as a measure of personhood, dignity, or moral perfectibility," by the reality of monozygotic twins and the possible occurrence of cloned persons in the future. It certainly represents "the legacy of individualism" and "an unrealistic view of individual autonomy." "Nature is more relational. DNA does not make a person all by itself. . . . [For] once the embryo attaches to the mother's uterine wall about the fourteenth day, it receives hormonal signals from the mother that precipitates the very gene expressions necessary for growth and development into a child."[28] Ontologically, relationality precedes innateness. "Dignity is first conferred relationally, and then it is claimed independently." Theologically, dignity is ultimately a gift conferred by the grace of God.

Furthermore, Christian personhood is a communitarian concept in the context of the doctrine of the Trinity, "persons-in-communion" or "persons-in-relation." "The self-relatedness of God makes possible the self-relatedness of human beings; the other-relatedness of God makes possible the other-relatedness of human beings."[29] This relatedness denotes an "openness of being" beyond the individual and his or her biological origin. "A person is a self in the process of transcending the boundaries of the self. This self-transcendence is the root of freedom. . . . True personhood arises through a trans-biological communion with God that transforms our relationship to the physical world."[30]

Underscoring Christian eschatology, the third model argues that the dignity of a person is "proleptic," or "future oriented."[31] Peters states:

> Dignity derives more from destiny than from origin, more from our future than from our past. . . . Persons whom we know and love today are on the way, so to speak; they anticipate their full essence as human beings by anticipating their resurrection and unity with Christ within the divine life. Our present dignity is itself part of this anticipation, a prolepsis of our eternal value conferred upon us by the eternal God. Dignity is not originally innate; it is eschatological and retroactively innate. . . . Our final dignity, from the point of view of the Christian faith, is eschatological; it accompanies our fulfillment of the image of God. Rather than something imparted with our genetic code or accompanying us when we are born, dignity is the future end product of God's saving grace

activity which anticipates socially when we confer dignity on those who do not yet claim it. The ethics of God's kingdom in our time and in our place consists of conferring dignity and inviting persons to claim dignity as a prolepsis of its future fulfillment.[32]

As Peters elaborated, Christian theological anthropology advocates the person in relationship and persons as prolepsis of the eschatological humanity. This refers to the interpretation of the doctrine of the image of God (*imago Dei*) fully manifested in the personhood of Jesus Christ.

## Respect for Life

Definitions of dignity so far defined are relatively anthropocentric and rather elitist by emphasizing full personhood. So another important language used in the discussion is *respect*. Ethical bodies such as the U.S. Human Embryo Research Panel (1994), the U.S. National Bioethics Advisory Commission (1999), and the Geron Ethics Advisory Committee (1999) declared that the human embryo (and so the blastocyst) should be treated with proper and appropriate respect. Although the embryo is not regarded as fully as an individual person, it is entitled to respect. However, the meaning of respect is "illusive." What does "respect" really mean in the context of hES cell research at the points of the creation, derivation, and manipulation of human embryos? To clarify this question, Karen Lebacqz suggested five models of respect in relation to persons, nonpersons, sentient beings, plants, and the ecosystem.[33]

*Respect for person* is again based on the Kantian criteria for personhood. It attributes the distinctiveness of personhood to the self-determination or autonomy, i.e., the ability to reason, to use the rational will, and to govern conduct by rules (*auto-nomos*). Respect in this context includes "active sympathy and readiness to hear the reasons of others and to consider that their rules might be valid."[34] However, embryos lack the ability of self-determination or autonomy, though they may have a potential to develop reason and become rule-governed beings. The Kantian personhood does not fit the nature of embryos.

The Judeo-Christian tradition endows a preferential option to the poor or to the *minjung* (民衆), the most vulnerable members of society to oppression, such as the outcast, strangers, sojourners, orphans, and widows. From this vantage point, though, embryos are not a Kantian person, "the requirement for respect is not diminished."[35] However, this second model is still anthropocentric. At this point, *respect for sentient beings* becomes relevant. Since sentience (an ability to feel pain and suffering) is distinctive of personhood, it can be the basis of respect for those who are not fully persons. However, Lebacqz does not solicit a full vegetarianism. Humans are permitted to slaughter animals for the sake of nourishment. Respect in this context refers to a requirement that pain, fear, and stress should be minimized. The biblical law that prohibits ingesting the blood of animals and the Native American practice of prayer to ask forgiveness of an animal to be killed for food illuminate this insight. The implication of this third model for the research is that the destruction or manipulation of embryos is not necessarily disrespectful but requires great care and commitment to minimizing pain and reducing fear or stress.

Neither is the early embryo, still without the physical capability for feeling and emotion, a sentient being. The minimization of pain and the reduction of fear are not so relevant in this context. Hence, the fourth and the fifth models concern nonsentient things such as plants and the ecosystem. Respect in these models implies attention to "the concrete reality" or "the independent value" of the other and of the ecosystem.[36] Respect requires us "to perceive the other in itself" and to see nature "as valuable *in and of itself*" rather than to see it as valuable *for us*. That necessitates decentering human perspective with deep epistemic humility and seeing other beings and things, including "not just sentient creatures, but land, rocks, trees, and rivers," "with respect and awe."[37] Lebacqz summarized the implications of this analysis with respect to the hES cell research.

Researchers show respect toward autonomous persons by engaging in careful practices of informed consent. They show respect toward sentient beings by limiting pain and fear. They can show respect toward early embryonic tissue by engaging in careful practices of research ethics that involve weighing the necessity of using this tissue, limiting the way it is to be handled and even spoken about, and honoring its potential to become a human person by choosing life over death where possible.[38]

### Some Preliminary East Asian Christian Reflections

Through the advance of embryology and genetics, the sanctity of life has become an ambiguous and illusive notion. Today, this concept is subject to reinterpretation that may necessitate a demythologization and even a scientification. This issue is one of the most significant and practical hermeneutical imperatives we are facing now. So far I have surveyed some major themes generated in North American discussions focusing on Christian theology and ethics. Overall, they do not seem to overcome fully their Western legacies of substantialism, individualism, and anthropocentricism (dignity, person, and respect as individual entity). Nonetheless, there are some intriguing developments to my East Asian Christian eyes. Here are some of my preliminary reflections on those.

1. The language of respect is especially fascinating, because *gyeong* (*ching*, 敬) stands at the heart of Korean Neo-Confucian thought culminated by Yi T'oegye (1501–1570). For T'oegye, *gyeong* entails not only epistemic humility but also profound ecological sensitivity (avoid treading the ant mounds!). *Gyeong* signifies the state of human mind-and-heart ready to realize its ontological psychosomatic union with nature, to fulfill the Confucian theanthropocosmic vision (the communion among the triad of Heaven, Earth, and humanity, 天地人). By attaining this state of mind, a person can possess an ability to hear the voices and feel the pain and sufferings of nature (commiseration) and can exercise beneficence (*jen*, 仁) whose attributes Confucianism calls Four Beginnings, namely, humanity, propriety, righteousness, and wisdom.[39] Western bioethicists such as Lebacqz are eagerly searching for precisely the state that Confucian sages saw as the starting point for self-cultivation in their students. In his doctrine of "the Oneness of All Things," Wang Yang-ming articulated:

> The great man [sic.] regards Heaven, Earth, and the myriad things as one body. He regards the world as one family and the country as one person. As those who make a cleavage between objects and distinguish between the self and others, they are small men [sic.]. That the great man [sic] can regard Heaven, Earth, and the myriad things as one body is not because he deliberately wants to do so, but because it is the natural humane nature of his mind that he does so.[40]

2. The definition of human personhood as Peters endeavored to introduce also resonates with the Neo-Confucian anthropology. The human person in Confucianism does not mean "a self-fulfilled, individual ego in the modern sense, but a communal self or the togetherness of a self as 'a center of relationship.'"[41] The second model of Peters, person-in-relation, is in fact the basic presupposition for the whole Confucian project. The crucial Confucian notion of *jen* denotes the ontology of humanity as being-in-relationship or being-in-togetherness. Again, in his famous passage of the Western Inscription, Chang Tsai (Jang Ja) wrote:

> Heaven is my father and Earth is my mother, and even such a small creature as I finds an intimate place in their midst. Therefore, that which fills the universe I regard as my body and that which directs the universe I consider as my nature. All people are my brothers and sisters, and all things are my companions. . . .[42]

Moreover, Confucianism regards humanity as the heavenly endowment (*T'ien-ming*, 天命), in the similar manner as Christian theology understands humanity as the image of God (*imago Dei*). The dialogue I developed between John Calvin and T'oegye shows clearly this characteristic of a relational and transcendental anthropology.

> The Christian doctrine of *Imago Dei* and the Neo-Confucian concept of *T'ien-ming* reveal saliently this characteristic of a relational and transcendental anthropology. Calvin and T'oegye are the same in defining humanity as a mirror or a microcosm to image and reflect the glory and the goodness of the transcendent ground of being.[43]

3. The final model of Peters, person as proleptic, is most intriguing. Dignity refers to *telos* (destiny) rather than origin. Peters brought the developmental and eschatological dimensions of Christian anthropol-

ogy to the stem-cell debates. Dignity of life in the Christian sense ultimately (*telos*) means the eschatological personhood of Jesus Christ, just as that in the Confucian sense denotes the futuristic sagehood (聖人). In fact, this is precisely the significance of the Christian doctrine of sanctification and the Confucian teaching of self-cultivation. The sanctity of life has been ontologically and eschatologically conferred (as it is the *T'ien-ming* and the *imago Dei*) in every stage; it is primarily given (relational) rather than innate (substantial). Yet, in existence, it is ambivalent because of this transcendental potentiality (transcendent yet immanent). So it requires a rigorous process of self-realization, i.e., sanctification and self-cultivation.

The human embryo should be treated with respect or reverence, but it is the sanctity of life in potential and in prolepsis that necessitates a full realization. The hES cell research can be a means to this self-realization of life (embryos) in the cosmic relationship. But life science cannot be the *telos* for itself in dealing with any stage or form of life system. Science is neither inevitable nor unstoppable.

4. From the East Asian Christian perspective, therefore, the sanctity of life rather implies the imperative for a life to realize itself to the fullest end of what it ought to be. This involves the diligent practice of sanctification and self-cultivation in mindfulness (or respect). A researcher in a laboratory is also a human person who needs to engage in this rigorous practice of mindfulness. This understanding may be the prerequisite to exercise one's freedom to help other forms of life accomplish their imperative for self-realization. But it denotes not so much a self-defined created cocreator self-consciously stretching to create techno-sapiens or the superhuman cyborg, as a humble cosmic co-sojourner participating in the great transformative movement of the anthropocosmic trajectory, i.e., the Tao (the Way).[44] Finally, the sanctity of life from an East Asian Christian perspective means a fulfillment and embodiment of the proleptic Tao, in its own freedom of life (*wu-wei*, 無爲). After all, both science and religion are taos for life, the great openness for cosmic vitality (*saeng-myeong*, 生命).

## Notes

1. Philip Hefner, "Biocultural Evolution and the Created Co-Creator," in *Science and Theology: The New Consonance*, edited by Ted Peters (Boulder, CO: Westview Press, 1998), 174–88.
2. A Japanese word that means "evocative question."
3. United Nations, Press Release GA/10333, "United Nations Declaration on Human Cloning," March 8, 2005, available at http://www.un.org/News/Press/docs/2005/ga10333.doc.htm.
4. United Nations, Press Release GA/L/3271, "Legal Committee Recommends UN Declaration on Human Cloning to General Assembly," February 18, 2005, available at http://www.un.org/News/Press/docs/2005/gal3271.doc.htm.
5. Thomas A. Shannon, "From the Micro to the Macro," in *The Human Embryonic Stem Cell Debate: Science, Ethics, and Public Policy*, edited by Suzanne Holland, Karen Lebacqz, and Laurie Zoloth (Cambridge, MA: MIT Press, 2001), 181.
6. Ibid.
7. Ibid., 182.
8. Ibid., 183.
9. Paul R. Wolpe and Glenn McGee, "'Expert Bioethics' as Professional Discourse: The Case of Stem Cells," *Human Embryonic Stem Cell Debate*, 185.
10. Ibid., 186.
11. Ibid., 187.
12. Ibid., 194–95. Italics are from the text.
13. Ibid., 189.
14. Ted Peters, ed., *Genetics* (Cleveland, OH: Pilgrim Press, 1998), 33.
15. The Division of Mission in Canada, "A Brief to the Royal Commission on New Reproductive Technologies on Behalf of the United Church of Canada," January 17, 1991, 14, cited in Peters, *Genetics*, 33.
16. Southern Baptist Convention, "On Human Embryonic and Stem Cell Research," available at http://www.sbcannualmeeting.org/sbc99/res7.htm, cited in Ted Peters, "Embryonic Persons in the Cloning and Stem Cell Debates," *Theology and Science* 1, no. 1 (2003): 58.
17. The General Board of Church and Society, "Letter to Extend Moratorium on Human Embryonic Stem Cell Research," from Jom Winker to President George W. Bush, July 17, 2001.
18. Pontifical Academy Life, "Declaration on the Production on the Scientific and Therapeutic Use of Human Embryonic Stem Cells," Vatican City, August 2000, cited in Peters, "Embryonic Persons," 59.
19. Ibid., 60.
20. Pope John Paul II, "Evolution and the Living God," in Peters, *Science and Theology*, 151.
21. Pope John Paul II. *Evangelium Vitae* (25 March). *Acta Apostolicae Sedis* 87 (1995): 401–522, cited from Peters, "Embryonic Persons," 59.
22. Richard Doerflinger, "The Policy and Politics of Embryonic Stem Cell

Research." *The National Catholic Bioethics Quarterly* 1:2 (Summer 2001): 143, cited from Peters, "Embryonic Persons," 59.

23. Peters, "Embryonic Persons," 64.
24. Ibid., 68.
25. Ibid., 68–72.
26. Ibid., 68.
27. Ibid., 69.
28. Ibid., 70.
29. Ibid., 70f.
30. Ibid., 71.
31. Ibid.
32. Ibid., 72.
33. Karen Lebacqz, "On the Elusive Nature of Respect," in *Human Embryonic Stem Cell Debate*, 149–62.
34. Ibid., 151.
35. Ibid., 153.
36. Ibid., 156–57.
37. Ibid., 156, 159.
38. Ibid., 160.
39. Michael C. Kalton, *To Become a Sage: The Ten Diagrams on Sage Learning by Yi T'oegye* (New York: Columbia University Press, 1988), 119–41.
40. Chan Wing-tsit, *Instructions for Practical Living and Other Neo-Confucian Writings* (New York: Columbia University Press, 1963), 272; also see Heup Young Kim, *Wang Yang-ming and Karl Barth: A Confucian-Christian Dialogue* (Lanham, MD: University of America Press, 1996), 43.
41. Heup Young Kim, *Christ and the Tao* (Hong Kong: Christian Conference of Asia, 2003), 12.
42. Chan Wing-tsit, *A Source Book in Chinese Philosophy* (Princeton, NJ: Princeton University Press, 1963), 497–98.
43. Kim, *Christ and the Tao*, 91.
44. Hefner has made an intriguing proposal of the created cocreator. "*Homo sapiens* is created co-creator, whose purpose is the stretching or enabling of the systems of nature so that they can participate in God's purposes in the mode of freedom. As a metaphor it describes the meaning of biocultural evolution and therefore contributes to our understanding of nature as a whole" (Hefner, "Biocultural Evolution," 174). Although sympathetic toward it, the East Asian Christian perspective does not endorse this proposal. For it still carries over the language of substantialism, individualism, and anthropocentricism as well as the modern paradigm of domination and control. In the East Asian Christian perspective, the freedom of life refers not so much to "the freedom to alter," change, or modify nature or life systems, but rather "the freedom not to alter" unless it is ultimately helpful for ecological and cosmic sanctification. It warns against the dangerous rise of techno eugenics under the pretext of superhuman ideology. Here, the distinction "between eugenic purposes and compassion

purposes" is well taken. Genetic selection to help prevent or reduce suffering may be understandable, but genetic engineering to enhance genetic potential to produce "designer babies" (imagine their quality control!) in "the perfect child syndrome" is not permissible (see Ted Peters, *Science, Theology, and Ethics* [Hants, England: Ashgate, 2003], 191–99).

## Bibliography

Chan Wing-tsit. *Instructions for Practical Living and Other Neo-Confucian Writings*. New York: Columbia University Press, 1963.

———. *A Source Book in Chinese Philosophy.* Princeton: Princeton University Press, 1963.

Holland, Suzanne, Karen Lebacqz, and Laurie Zoloth. Eds. *The Human Embryonic Stem Cell Debate: Science, Ethics, and Public Policy.* Cambridge, MA, London: MIT Press, 2001.

Kalton, Michael C. *To Become a Sage: The Ten Diagrams on Sage Learning by Yi T'oegye.* New York: Columbia University Press, 1988.

Kim Heup Young. *Wang Yang-ming and Karl Barth: A Confucian-Christian Dialogue.* Lanham, MD: University of America Press, 1996.

———. *Christ and the Tao.* Hong Kong: Christian Conference of Asia, 2003.

Peters, Ted. "Embryonic Persons in the Cloning and Stem Cell Debates," *Theology and Science*, 1/1 (2003), 51–77.

———. Ed. *Genetics.* Cleveland: Pilgrim Press, 1998.

———. Ed. *Science and Theology: The New Consonance.* Boulder, CO: Westview Press, 1998.

# Editor's Introduction to Aut Moses, Aut Darwin?

**A. Markoš, F. Grygar, L. Hajnal, K. Kleisner, Z. Kratochvíl, and Z. Neubauer,** Charles University, Czech Republic

The team headed by Anton Markoš at Charles University in Prague is committed to reinvesting life and the changing universe with intrinsic specialness, a quality too casually stripped away by the vulgar objectification of simple rationalist logics. Evolution, to them, is an ineffable experience that can only be fully understood as a participatory process, knowable only through the personal acquaintance by which our minds become "coevals, successors, shareholders and heirs of parallel life courses, historical trends, and cosmic events." Keying off on the growing public discussion that seeks to pit Darwinism against creationism, Markoš et al. present a deep philosophical argument for an alternative view of the unfolding of life on Earth.

Markoš et al. were inspired in this essay by a questionnaire on evolutionary theory circulated in the Czech Republic in 2005. Its organizers sought detailed responses from the country's intellectual establishment but framed it in a polar form by pitting narrowly construed neo-Darwinism against vulgar forms of creationism. The authors undertake to show that this is a false dichotomy whose origins lie in the history of evolutionary thinking and one that can be overcome through careful attention to the philosophical consequences of a dynamic, changeful living world.

The central misunderstanding that shrouds mechanistic approaches to evolution is their failure to encompass the activeness, the participatory "thingness" (in the Heideggerian sense) of living entities. Life's urgent, developmental nature demands that it seek self- and other-knowledge and engenders the motif of "struggle" that infuses evolutionary thought. But the word "struggle" has been distorted in the evolutionary dialectic by an overemphasis on warlike metaphors, metaphors that have likewise come to dominate the metaconversation between adherents of evolution and creation.

In fact, rational constructs, "theory" in the locution of modern science, and *theoria*, the ancient idea of spectacle, are both at play in the construction of evolution. Markoš et al. point out that no rational construct can contain its own origin and so any point of creation must, by definition, live outside a "theory" in the scientific sense. They also note that biblical creation stories embrace a rational wholeness creating an ecological *mise-en-scene* for the narrative of cosmic unfolding. And, on the scientific side, they see a tension between holistic, narrative, and historical perspectives on the one hand and a reduction to quantifiable data, specification, and logical abstraction on the other.

Ultimately, all life exists in a state of negotiation with itself and its ecological partners. And the authors assert that a biosemiotic acceptance of the codes that span levels—both in and between ecosystems—offers a substantial new base on which to reconsider the evolution of life. Communication and the codes through which communications are formed, from sexual reproduction to predation and competition for resources, "represent facets, integral parts of the bodily existence of living beings, beings who *care* about their being and who maintain uninterrupted corporeal lineages from the very beginnings of life on our planet."

P.D.

## 8

# Aut Moses, Aut Darwin?

**A. Markoš, F. Grygar, L. Hajnal, K. Kleisner, Z. Kratochvíl, and Z. Neubauer,** Charles University, Czech Republic

> Thinging, the thing stays the united four, earth and sky, divinities and mortals, in the simple onefold of their self-united fourfold. . . . This appropriating mirror-play of the simple onefold of earth and sky, divinities and mortals, we call the world. The world presences by worlding. That means: the world's worlding cannot be explained by anything else nor can it be fathomed through anything else.
>
> Martin Heidegger

> It is a stunning fact that the universe has given rise to entities that do, daily, modify the universe to their own ends. We shall call this capacity *agency.*
>
> Stuart A. Kauffman and Philip Clayton

Evolutionary teaching began as an attempt to return to the Renaissance ideal of knowledge: by turning attention away from what is general and typical, toward what is natural, unique, and individual, and by replacing measurement with observation and comparison; hence restoring confidence in both individual senses and personal experience. Evolutionary teaching rediscovered the importance of history—long forgotten since the Renaissance.

This attempt to introduce evolutionary (i.e., natural) thinking as an antinomy of rationalism took place, however, during the hegemony of modern scientific rationality, when "reality" was exclusively understood as "objective." In such an atmosphere, evolutionary teaching also claimed the status of objective knowledge, even when it factually challenged it; solely by assenting to such a compromise could it be allowed to establish itself as a science. The compromise, however, prevented evolutionary science from being fully consequent, from broadening the evolutionary approach to the whole of reality: it was unthinkable at the time, that not only living nature, but the entire reality, even our knowledge and truth—even science and evolutionary science!—is subject to evolution,[1] that the history of the world means narratives with an open end. It was unthinkable to state that the evolution of knowledge (with science being no exception) does not approach some definite level of understanding, that it does not even recognize any preferred direction toward it. It would have been a scandal to announce that the directions, goals, topics, and motifs of acquiring knowledge become redefined again and again with every new piece of knowledge, theory, or teaching, and that the horizons are floating with every newly formulated truth; to admit that the evolution of a species is but a special case of evolution, wherein every newly appearing individual opens the field of possibilities differently, and often in new directions.

This regrettable atmosphere governs our thinking even today. Even today, for far too many, the consequences of an evolutionary approach appear to be too big a chunk to swallow;[2] and for proponents of traditional science it remains fully indigestible, because it obviously denies the very core of objectivity. This may be the reason why, to our knowledge, none of the founders of evolutionary teaching drew such consequences. In contrast with the nineteenth century, the contemporary alternative movements mentioned above have much better chances to succeed.

In biology, the evolutionary approach weakens continually nowadays (in spite of the lip service paid to it under the banners of the neo-Darwinian genocentric revolution), paradoxically, just as the whole of nature—ever inorganic—falls under its spell! Today, mass itself, the whole of the universe, is viewed by physics as resulting from long and complicated evolutionary processes. The same holds for our own history, including the history of religion—and also of science.

As a consequence, the scientific conception of reality as objective ceases to be the self-evident, uncritical, and undeflected reference point of human thinking. New developments and discoveries need not be interpreted exclusively in the light of a single, scientific rationality. They can therefore refer also to other traditions of knowledge, e.g., Christian or hermetic; they can return to the origins of modern science and follow alternative directions that have become forgotten. They can also become part of cultural experience, understood more deeply. Above all, however, this favorable atmosphere allows a rehabilitation and reevaluation of evolutionary thinking, whose advent was ahead of its time.

What became fatal for Darwinism and its derivatives was their obstinate effort to express themselves as orthodox science. Mechanistic determinism and reductionism, attitudes sincerely but groundlessly declared in most cases, led to their transformation into "neo-Darwinism" (i.e., neo-Darwinism of the *fin de 19ᵉ siècle*, not its extant form), which entailed a return to the formalism of objective knowledge (especially thanks to Weismann). Its hylozoic materialism—incompatible with mechanicism—its atheism and positivism—bound to materialism—have finally declined to mere declarations (the clash with religion being completed in that time).

### The Controversy around Darwin as a Symptom

The emergence of any particular form, i.e., universe, life, man, etc., is impossible to grasp, i.e., understand or deduce, via logos (*ratio*). Logos is even unable to express novelty, let alone explain or explicate it; it is based in relationship, but being and nonbeing cannot be commensurated, put in ratio, rationalized. The very essence of beginning and becoming (*physis*, *natura*), as well as *all development in time is, in this respect, irrational.* Every conception is "virginal," untouched by any descent or intention, without bearing any burden of *cause* (*aitia*). It simply happened, by chance, freely, spontaneously, without any cause, "just so." From nothing there arose something, standing for itself, self-confident, natural. The beginning of what exists naturally (*to fysei on*) is the foundation of any creative action unwinding from inside, of self-formation.

Nothing of what arises, what comes up by simple appearing and becoming, can fail to enter into the content of knowledge (as logos), especially in the light of scientific rationality and objectivity. "What has been will be again, what has been done will be done again; there is nothing new under the sun," says the prophet (Eccl. 1:9). Beginning and conception are not items, objects, articles lying perfectly as *facts* in the world; their nature concerns subjects, stimulants, events—furnishing chances for the implementation of something that may prove its fitness. Moreover, novelty is principally unrivalled, unrepeatable, and unprecedented—as any natural thing. But *de singularibus non est scientia*, as goes the age-old precept. True, exact science is not casuistic, nor is it a factography. The true subject of science is what can be generalized, put under some rule or law: it is knowledge of things having "objective reality," i.e., it concerns objects (*obiectum*). Scientific knowledge is an endeavor focused on the causes of whole categories of things, intended to reveal how things are "in reality." Therefore, it must first transform a thing into an object because only objects can become the specific material, the "subject," of knowledge. This rational base also provides an explication of the things themselves, their *raîson d'être*—it explains why the things exist at all and what essence is represented by this very exemplar. Only objective reality makes such an item real, because an absolute and mandatory knowledge is possible only with objective reality.

## Things and Objects

"Science always encounters only what *its* kind of representation has admitted beforehand as an object possible for science," says Heidegger in his famous essay "Thing" (1971, 168), and continues: "Science's knowledge, which is compelling within its own sphere, the sphere of objects, already has annihilated things as things long before the atom bomb exploded. . . . The thingness of the thing remains concealed, forgotten."

What, then, is "the nature of the thing," what is "thinghood," and what is "thinging of the thing" and "worlding of the world," topics developed further in Heidegger's essay? The thing is part of the world, its affairs and contexts, part of its "presencing," i.e., the emanating or revealing of its presence. It is a symbol, it takes on meaning in an everlasting multicontextual game. Things are part of our lives, but objects are not: objectivization kills life; "objective reality" hasn't the essence of notions like *world*, *development*, *life*, *meaning*, or *emergence*. No naturally existing things exist *objectively*, therefore they are not available to exact scientific knowledge. At the same time, *natural experience* concerns exactly such natural things, actions or events transformed into familiar knowledge or understanding. Even the very process of acquiring understanding (including the objective), i.e., the emergence of truth, acquaintance with the world and with the self, is of such nature. Such individual experience is not an *ob*ject of understanding, but its *sub*ject. Knowledge is part of the content of such an experience and vice versa—experience, not objective reality, lies in the background of knowledge. In other words, our natural experience does not consist of data and knowledge, but ideas and events based in imagination and narratives. We think in stories (Gregory Bateson), in myths, and their stuff consists of likenesses, analogies, archetypes, and symbols. Natural reality is of the same nature (stuff)—she is not something to be accepted as ready-given in textbooks. She is *the presentation*, and she gains by passing, by tradition: stories beget more stories. We also participate in *this* world—we are as real as the embodied experience we incessantly seize, continuously understanding it anew and in a new likeness. Evolution—unfolding from inside of the self—is therefore a basic, natural way of our being-in-the-world; it is the "*a priori* form" of our experiencing reality, in which we ourselves are given (defined) as its partial subject.

Hence, evolution looks elusive and counterintuitive, she resists commonplace understanding because she is not the *content* of knowledge, but a way in which knowledge is acquired, how it becomes content. Because of this, evolutionary thinking is very hard to transform into reflected cognizance—it is rather an attendant awareness that enables awakening and reflection. Evolution—the course of metamorphoses, reality *in statu nascendi*—is an experience that cannot be transmitted, only participated in: it is not based on common knowledge, but on personal acquaintance. The evolutionary dimension of reality cannot be made a matter of common consciousness—rather it enables our separate minds to be partners and mutual witnesses in shared knowledge and in the adventure of discovery. In this way we become coevals, successors, shareholders, and heirs of parallel life courses, historical trends, and cosmic events. From this the tissue of the *world* is woven as a common dwelling place of related, intimate critters, naturally kindred by their origin, life stories, communication, and cooperation. The natural, evolving world thus exists as a commonly acquired experience of individuals within nature. The world is not a framework; it is neither a source nor a goal of an absolute, i.e., objective and indifferent knowledge; rather it is the vanishing point of the process of relative, contextual knowledge. The natural world is built of bounds, encounters, and commitments unifying her dwellers as cells are unified in living tissue.

### The Contest of Likenesses as a Manifestation of Will to Power, i.e., the Struggle for Life

The quest after the origin of the universe and of life is a streaming toward verity, natural genuineness. Natural truth is urgently pressing toward realization—its implementation is binding. Its nature is a Nietzschean will for power: "Life must again and again surpass itself."[3] Life is not an object, but a projection, embodied intentionality. To live means *to develop*, i.e., to expend an effort, to break some resistance: life is development, evolution[4] of its own powers, potential struggling to come up, get through, excel. This is the very source of the "struggle for life"—the mutual contest and competition of living forms.

A short digression: for biologists who are not native speakers of English, it always comes as a surprise when they happen to look for the entry of the word "struggle" in an English explanatory dictionary. This is because probably all European languages translate "struggle" (in the expression "struggle for life") unequivocally as "fight, battle, or combat."[5] Indeed, in Webster's Collegiate (1991) we read: "1: Contest, *strife*; 2: a violent effort or exertion: an act of strongly motivated *striving*." The entry "to strive," in its turn, informs us that the original meaning was "to endeavor."

Such warlike translations (prescribed, of course, by the exegetic tradition established in the late nineteenth century) may distort the whole understanding toward a single meaning. We therefore tried to read "struggle for life" as "endeavor to live"—we encourage the reader to do so as well—and we got a very peculiar insight into origins of species.

Now back to our topic: according to Heraclitus, "war (conflict) is the father of all things" (*polemos pater panton*), i.e., this also holds for all forms of life struggles as represented in various races. Each *species* (in Greek: *idea-eidos*) is defined by its specific appearance (again *idea-eidos*). A living being is thus an embodied likeness (*eidos*) of a specific "survival strategy" driven by many generations of predecessors; a successful strategy indeed as it survives to our day.

This is how Darwin should be read: his *On the Origin of Species* does not treat the question of the *birth* of species, but their descent or lineage of bodily appearance, i.e., forms of living. Their diversity is but a secondary *by-product* of repeated selection of those forms that endeavored and succeeded more often than others to propagate their generic traits into their progeny, i.e., to inform this progeny about *their* version of body likeness. Each living form represents a specific relationship with the world, a surviving strategy, that both stretches out toward the horizon of the species, and focuses in to the center, toward its likeness.

## Moses or Darwin?

What is the philosophical background of the "battle" between supporters of the Darwinian view, and adherents of the creation "theory," that derives the origin of life and the universe from an act of biblical creation? *Theoria* meant originally a spectacle, i.e., an unengaged way of looking; it referred to the particular perspective that was open for an audience—therefore "theory" in the sense of view, approach, or attitude. In science, "theory" received the meaning *rational construct*, a tool of systematic explanation, or a manual on how to identify and classify the phenomena in question—how to differentiate or amalgamate them according to theoretical, i.e., general and external, criteria, and how to put them into mutual relations, links, and dependences. *Theoria*, originally a vista, a view, a way of looking and observing, has turned into "theory": a set of instructions on how to systematize work, along with compilations of facts collected into proportions and relations so as to bind said facts into a closed, rational construction.

If we take this last meaning of the word, then, of course, no sovereign, divine "act of creation" can ever become a foundation for, or even part of, any theory. Reality can be regarded as the World of God—as a manifestation of his wisdom, goodness, and sovereignty. One can marvel at the performance, admire God's work, and praise him. However, from the "fact of creation" no other fact can be derived, nor is it possible to insert external facts into it. Creation provides the auditorium, not the starting point: neither the emergence nor the origin of anything else can follow. It is characteristic that the scriptures, when mentioning the three acts of biblical creation (Heb. *bara*—Gen. 1:1; 1:21; 1:26), never speak about the origin of the "world" or of "life," let alone give them a rational explanation.[6] Such motifs, notions, questions, and problems are anachronic, raising inadequate claims. Once again: no rational theory is able to explain origin and descent, because beginnings cannot be—*ex definitione*—included within any kind of relationships.

Creationism, however, does justify its way of knowing the world with the help of rational theories. It does so because it is a theory in a sense of "view": it takes reality as something that is here in its wholeness, closed within itself and ready-made; also made deprived of its origin, its ends, as well as similar discontinuities, i.e., gaps in reasoning. The "act of creation" is truly a negation of descent; it questions the naturalness of the world, and requires it to be rational.

But biblical revelation does not correspond to such demands. It contains something like an "ecological" layout of reality. Instead of narration, it provides mere exposition of stage settings, the scenery for future stories. The plot is the world itself, with its evolution in the form of an "evolutionary game" between man and God, between the Chosen People and other nations, and between individuals—"creation" thus denotes humankind, along with every human being as well as the cosmological event.

While the biblical entry provides a sketch, a framework of future events, evolutionary teaching comes with an explanation of the present situation as an outcome of events past. Its very content is a myth disguised as a scientific explanation. As convincingly shown by E. Rádl (1905–1913, excerpt in English, 1930), Darwinism is by its nature *irrational*; it represents a protest against the rationalism of traditional science.

Rádl referred to the crisis of Darwinism and predicted its quick decline. This was not, as it may seem, an unfortunate mistake, nor a manifestation of Rádl's prejudices, but a result of a thorough analysis and diagnosis of contemporary Weismannian neo-Darwinism as a factual denial of Darwinian reformation. Original Darwinism tried to transform biology into a historical science, whereas neo-Darwinism is a retrogressive rationalization, a rechanneling of evolutionary teaching back to the realm of objective science. This trend has continued throughout the whole of the twentieth century: biology today, built on molecular genetics, has totally ceased to be a science about living reality and has lost its connection with any kind of natural experience. Today, neo-Darwinism *is* rational indeed—at the price of forgetfulness of nature (a Heideggerian term).

Darwinism was built on the empirical self-stylization of science. However, science is not empirical, but experimental: its knowledge is not based in experience, but lies in attempts to transform phenomena into data that will fit into a rational framework or scheme, to fulfill theoretical (in the second meaning of "theory") expectations. Scientific observations are *supposed* to satisfy empirical methods, but only when we understand them in this way. Therefore they focus their attention on phenomena that can be expressed as numbers. The advantage of such an "empiry" is the possibility of generating mutually comparative data (independent of what is observed), of transmitting said data (independent of the observer), and publishing it (independent of knowledge). Thus, empirical *data* can be both easily conveyed and applied, independently of experience or knowledge.

In contrast, evolutionary knowledge is rather an enquiry. It is oriented symptomatically not systematically; its proceedings are heuristic and exegetic—revealing and explaining. To be able to understand what is going on now, it is interested in a concrete plot of what took place in the past. Thus, reconstruction of the past allows understanding of the present; to succeed in such an endeavor it is necessary to interpret present phenomena as witnesses, traces, indications, remnants, and echoes of life in the past. The question "Why are things so-and-so?" has, for an evolutionist, a simple answer: "Because it happened like this!"

Legitimate objections will immediately arise as, of course, this is not science! It isn't; we should, however, immediately add that this is the very way in which natural experience evolves: it also ignores inner causes, deeper reasons, faraway goals, and higher perspectives, and sticks to general feelings and impressions left by the past, revealing the present, and laying contours of the future. In antiquity, such thinking was labeled *doxa,* as likeness was attached to form because it remained with the phenomena but did not penetrate to its essence; this was contrasted with "true" knowledge, *episteme.* It is obvious that knowledge of this sort does not require as rigorous a demand on the form of data as it does in the case of scientific empiricism. In this respect, evolutionism was very accommodating toward biologists, because neither biological knowledge nor its data matches the criteria of objectivity anyway.

Evolutionary teaching can be considered narrative to a much greater extent than biblical revelation. Modern (natural) science came on stage at a time when Christianity was already unable to unite Christians—neither by mutual agreement nor by force—on a single interpretation of the scriptures. Hence, after the Thirty Years' War the expectations of Europe became fixated on another text—the Galilean "Book of Nature"—in the hope that its text would not require any interpretation: it is "open," i.e., accessible to everybody and readable by anybody, *clare et distincte.* Moreover, this is the only book that is singular and common to all: the world as shared, rational, and understandable—

self-explanatory and understandable in itself (*interpres sui*). What is written is also given: the "text" is not only *about* reality, it *is* that very reality. For Nature to be a text requires that she does not exist "just so," she must represent a rational whole: she must represent a Whole, constructed (written, prescribed) by a rational Creator—the "Author" of the book that is Nature. Around this time the Creation became perceived as a system (construct), and the Creator turned into an Engineer (a Great Watchmaker at that time, a Programmer today). This blasphemous, if ancient, image of the Creator as a Producer (demiurge) is bound to the forgetting Nature: the world became commonplace—free of contradiction, unchangeable, exact, accurate, and sharp, fully transparent and present here and now. In short, it was given "objectivity."

It quickly turned out, however, that "reading" the book of Creation is (at least) as difficult a task as interpreting the book of Revelation. Clear images of reality "as it is," mediated by science, became more and more complicated and confused, in no way simpler than the previous exegesis of God's secrets. Its supposed unity was only rational, formal, abstract, based in mathematics and logic—such a unity is of course only spiritual, i.e., invisible and impossible to demonstrate. It supplied a form of "worldview," but no orientation whatsoever in the world. Moreover, with the progress of knowledge, the proclaimed unity became more and more obscured and unavailable, even for specialists, let alone laypeople. Again, as in the case of theology, what remained was a mere belief! In Science we trust!

Evolutionary teaching ought to have remedied this dreary state of affairs. A historical perspective of nature enriched the scientific worldview by adding the dimension of time. As the spatial dimension of the Copernican system restored the order in the celestial realm, so did the Darwinian perspective introduce a kind of order into natural manifoldness. But with it came also a new perspective, different from that offered by the scientific worldview: the evolutionary view introduced a much deeper turnaround than did the famous Copernican revolution. Albeit repeating the mantra of objective and rational approach to reality, the evolutionary view deprived the world of whatever architecture it might have previously had! It endangered the very subject of the scientific ideal, which consisted of a quest for exact and definite, logical and deductive knowledge. An exact, nomothetic natural science was to be transformed into an idiographic, casuistic historical science, dwelling in, and dealing with, particulars, contingencies, unrepeatable events explainable and explicable only from what happened before. And in the future, such explanations (even of our own existence!) were to replace objective explications based in theoretical constructions mirroring eternal laws, then in danger of yielding to historical reconstructions, to narratives that could make the present understandable.

## Communication

Life, at all levels of description, is a communicative structure. As a consequence, ontogeny, as well as evolution, results from an interpretative effort of life itself, and is not driven by external forces or preprogramming. This premise will serve as opposition toward two common, rational understandings of the living: as embodiment of eternal principles, or as interplay of an external environment with a genetic code. If the term *information* means more than a measurable quantum of bits, it cannot be otherwise. The biological playground looks cohesive thanks only to a great effort by the community of biologists; life itself, undisturbed, plays quite another game not constrained by the corrals they build. Life rests in a common undercurrent of meaningful communication among living beings, i.e., in their cultural, eidetic dimension. To illustrate our thesis, we take three examples of life's approaches:

1. The bacterial world in the Gaian perspective, as a single planetary being (Markoš 2002; Markoš et al. 2009).
2. The centuries-old problem of homology in multicellular eukaryotes. Are living beings similar because they "crystallize" according to some common principles, as supposed by old morphologists? Or is their similarity rooted in a shared predecessor? And if no such founder exists, then is similarity mere chance (homoplasy)? Or does imitation belong to the basic characteristics of life?

3. Finally, the problem that can be epitomized by the slogan evo-devo. After all, bodily structures trace their forebearers, not in ancestral structures, but simply in undifferentiated unicellular germs (zygotes or spores)! What is inherited are not structures, but capability, the power to build them. Where does this power reside, how is it transmitted?

All three layouts point toward a precarious realm where *meaning* dwells. We would like to show that understanding of the living is impossible before we succeed to smuggle—somehow—this concept into biology. As a prerequisite to this, we believe, first, that understanding one's own condition is the very prerequisite of being alive. Second, it is necessary that a living being is able to distinguish its partners in the biosphere and to establish a contact with them.

We tried to coin the notion of hermeneutics done *by* the living, as compared to approaching living beings as mechanical devices that do not care at all about their existence (Markoš 2002; Markoš et al 2009). We would like to demonstrate incessant symbolic ("horizontal") communication networks at different ecological levels, starting with cytoplasm viewed as an ecosystem of proteins, through multicellular bodies of various complexities and origins, and various kinds of symbioses ending at the level of planetary communication network.

Semiotics, as hermeneutics, is concerned with extracting meaning. With *bio*semiotics, immediately a question arises: *who* is it that understands in this case and how? The most obvious answer, "living beings," is not satisfactory; it is the very "null hypothesis" that should be proved or rejected. Such an answer would sound very suspicious anyway in contemporary biology rooted firmly in the "laws of physics and chemistry," where, as we have already seen, the basic property of objects studied by science is *being lifeless*, inert, submitted to rules given from outside.

M. Barbieri (2003) dared to raise conceptually and lexically acceptable scaffolding for such a biosemiotic discourse. A code can be defined as a set of rules that establishes a correspondence between two independent worlds (e.g., the Morse code, the Highway Code, and even very technically specific language). Such a set of rules cannot be implied from the material (physical, chemical, connectional) properties of the system: it must be either negotiated within that system, or imposed from outside. As no external conscious agency issuing the code is conceivable (the rule-giver, as is the Department of Transport in the case of traffic codes), we are left with internal settings. Accepting the existence of true codes (in addition to matter and information), i.e., rules given by historical conventions, gives biology a new reliable basis, a platform for putting known things into a new perspective.

It should be stressed that there are no virtual rules given in advance to the domains emerging in the evolution of life—such rules are being negotiated *within* such individual domains, and the necessary preconditions for their existence is memory, settled codes, historical experience, ways of processing this or that piece of information, interpretation of newly occurring data and situation, etc. Whereas domains defined by scientific emergentism (flames, gyres, clouds, stars, galaxies, etc.) will pop up (with such-and-such probability) whenever favorable conditions are met for their coming into existence, phenomena like cells, viruses, mammals, or antibiotic resistance are results of genuine evolution, and their coming into existence was a question of changing the very household of the universe. They represent singularities that changed the "state space" and could not be foreseen even in principle, not to speak about statistics (see Kauffman 2000; Kauffman and Clayton 2006). We can, and do, model their behavior with physical and emergentist approaches, and can learn a lot from this study, but we are not able to *create* them de novo.

Barbieri erected a solid platform—within biological sciences—that allows us to approach some problems previously not allowed in biology. According to him, the platform could have evolved by bottom-up evolution, from simple molecules. Yet we consider the result—agents moving on the platform—to be something like robots or some species of zombie, rather than genuine living beings. Beings dwelling in this realm resemble Heideggerian (1995) "world-impoverished" creatures. The world is withheld from them—they are simply driven, taken, and captivated by forces that are forever foreign to them; they can never truly comprehend the world. Their communication is not speech, because "to speak to one another means: to say something, show something to one another, and to entrust to one another mutually to what is shown. To speak with one another means: to tell something jointly, to show to one another what that which is claimed in the speaking says in the speaking, and what it, of itself, brings to light" (Heidegger 1982, 122).

We would like to see living beings *speak*: we chose, therefore, a top-down approach starting in hermeneutics, and we hope to land safely on Barbieri's platform and thus connect both realms of knowledge—those of science and of meaning. We propose that meaning, evolution, morphogenesis, imitation, mimicry, and pattern recognition, as well as understanding signals, patterns, or symbols found in other beings, and all the ways that lead evolution into new dimensions, creative inventing novelties, etc., represent facets, integral parts of the bodily existence of living beings, beings who *care* about their being, and who maintain uninterrupted corporeal lineages from the very beginnings of life on our planet. They are uniting the extant biosphere into a single, dynamic, semiotic space, which is kept together by the mutual interactions and experiences of all its extant inhabitants. Hence, the *codes sensu Barbieri* are, in our view, negotiated "from above," from shared language(s), merely as a useful tool that automatizes activities that can then be relied on, that need not be constantly renegotiated.

Negotiation means communication with . . . with whom? With sexual partners, with bacteria in our alimentary tract, with colleagues abroad. . . . Communication allows equivocality, the defocusing of all phenomena and all forms of reference to them, to precipitate from the field of possibilities: sometimes as well-entrenched patterns, sometimes as genuine novelties. At the same time, the existence of this superposed and commonly shared field allows mutual games of understanding, misunderstanding, cheating, and imitation at all levels of biosphere, e.g., the precipitation of actual versions of the fit—in Darwin's usage of the word.

Incessant creation is an engine of evolution.

## Notes

1. With the exception of thinkers like C. S. Peirce, F. Nietzsche, or H. Bergson—but they by no means belonged to the mainstream.
2. See how modern authors (e.g., Kauffman 2000; Barbieri 2003) struggle when attempting to reconcile the nature of life with scientific paradigms.
3. F. Nietzsche, *Thus Spake Zarathustra II* (Cambridge: Cambridge University Press, 2006), 29.
4. Expressions like *development* or *evolution* are quite clumsy and testify to the poverty of European languages originating from our millennial forgetfulness of nature. Such expressions clearly evoke an idea of unwinding something that has been waiting, up to now, rolled up, scrolled, hidden in a bud; of coming up with something preexisting that was *already and always there*, only hidden in its own levels. Recall that the teaching of preformism bore, in its heyday, the name "evolutionism"—only after Darwin did the word evolution receive a meaning quite opposite the original one.
5. German *Kampf ums Dasein*; Czech *boj o život*; Russian *bor'ba za zhizniu*; French *lutte pour la vie*; Hungarian *harc életre és halra*; Italian *lotta per la vita*; etc.
6. Biblically, "world" is called "age" (*olam*); "world," or even "creation of world" is nowhere to be found—at least not in Genesis 1–3.

## Bibliography


Barbieri, M. *The Organic Codes: An Introduction to Semantic Biology*. Cambridge: Cambridge University Press, 2003.

Heidegger, M. *The Fundamental Concepts of Metaphysics*. Bloomington: Indiana University Press, 1995.

———. "The Thing." In *Poetry, Language, Thought*. Translated by Albert Hofstadter. San Francisco: Harper, 1971, 161–84.

———. "The Way to Language." In *On the Way to Language*. Translated by Albert Hofstadter. San Francisco: Harper, 1982, 111–36.

Kauffman, S. A. *Investigations*. Oxford: Oxford University Press, 2000.

———, and P. Clayton. "On Emergence, Agency, and Organization." *Biology and Philosophy* 21 (2006): 501–21.

Markoš, A. "The Ontogeny of Gaia: The Role of Microorganisms in Planetary Information Network." *J. Theor. Biol.* 176 (1995): 175–80.

———. *Readers of the Book of Life*. Oxford: Oxford University Press, 2002.

Neubauer, Z. "Spor o Darwina jako Symbor" ["The Strait around Darwin as a Symbol"]. *Prostor* 65–66 (2005): 115–32.

Rádl, E. *Geschichte der Biologischen Theorien. I.* [*History of Biological Theories*] 2nd ed. Leipzig: Engelmann, 1913.

———. *Geschichte der Biologischen Theorien. II.* [*History of Biological Theories: Evolutionary Theories of Nineteenth Century*] Leipzig: Engelmann, 1909.

———. *The History of Biological Theories*. London: Oxford University Press, 1930.

Editor's Introduction to

# Human Origins: Continuous Evolution versus Punctual Creation

**Grzegorz Bugajak and Jacek Tomczyk,** Cardinal Stefan Wyszynski University, Poland

Bugajak and Tomczyk confront the prevalent dichotomy between a supposedly continuous process of evolution and a sudden, "punctual" divine creation. While they acknowledge that there are only blurry distinctions between humans and our evolutionary antecedents, they nonetheless argue that these distinctions are real and that humanity is unique among animals. They also deny the validity of creation stories that situate the entirety of the divine creative act at one point in time. Between these two, they find a space for nonopposition and hope for a "reconciliation of theological and scientific perspectives on the origin of man."

The authors begin their exploration with one of the most foundational and vexing problems in evolutionary theory—the validity and definition of the concept of species. Without a deeply meaningful construct of species, it is, of course, impossible to assert that humans are a "really new" kind of creature. Paleontologists wrestle with this challenge to an even greater extent than evolutionary biologists partly because they lack the capacity to directly observe or experiment upon the reproductive success of individuals and subpopulations. Genomic analysis, a powerful new tool in evolutionary studies, nonetheless lacks any natural demarcation between interspecial and interindividual genetic variation, leaving researchers to establish *essentially arbitrary* boundaries. Anthropologists likewise are subject to method-dependent definitions in their attempts to establish the borderlines between species as they sort through morphological, archeological, and genealogical information. In every case, science is left with *ad hoc* boundaries and lacks absolute, natural dividing lines with which to incontrovertibly define species.

Nonetheless, Bugajak and Tomczyk hold that there are evident gaps between humans and other animals, that a "continuity marked with leaps" has taken us to a substantially different level. Ethologists have established the presence of cultural elements in primates, for example, but certain advanced behaviors and characteristics are unique to humanity. Syntactic speech and art, the capacity for persistent hatred and our very broad altruism seem to the authors to separate us from the rest of the animal kingdom. Most importantly, our intellectual capacities enable us to conceptualize ideas as referents and to argue logically. As Aristotle held, it is this "pure knowledge" that is most desirable "because only it is worthy of man."

A belief in divine creation is often juxtaposed with evolutionary theory on the basis of a naïve misunderstanding. The authors argue that most nontheologians misapprehend both the temporal conundrum of cosmic creation and the complexities of human emergence. Citing fundamental Christian doctrine, they hold that "God created the world not *in* time but *with* time." Hence, it is no contradiction to allow for a fundamentally temporal process such as evolution to have always existed. There was no time "before" creation in which it would not have been. Since Pope Pius' 1950 encyclical, the Catholic theological door has been open to "a new understanding of the dogmatic truth of original sin." Humanization may have come about by evolution and monogenism's repudiation is resolvable by a focus on the polygenetic emergence of sinning human beings. The sophisticated tools of theological analysis have all too often been laid down in the evolution/creation controversy and the authors encourage us to reject the "false view of God who is external to the world."

If, as the authors argue, the oppositional appearance of evolution and creation is softened by closer analysis, a number of possible approaches present themselves. Calling upon the philosophical literature, Bugajak and Tomczyk offer a worldview that bridges the apparent conflict. They suggest adopting two primary assumptions: (1) that process is the primary reality of life and (2) that our cognitive abilities are "in accordance with the world" and reflect the correct action of our beings in creation. Although the latter assumption opens a door to affirming the essential rightness of spirituality (an evolved and, hence, appropriate expression of human being), the authors note that other approaches might be equally productive. Above all, they hope that by deconstructing the rigid barriers of opposition between the evolutionary and creation worldviews, a rich and productive new vision can emerge.

P.D.

# 9

# Human Origins

## Continuous Evolution versus Punctual Creation

**Grzegorz Bugajak and Jacek Tomczyk,** Cardinal Stefan Wyszynski University, Poland

### Introduction

One of the particular problems in the debate between science and theology regarding human origins seems to be an apparent controversy between continuous character of evolutionary processes leading to the origin of *Homo sapiens* and punctual understanding of the act of creation of man seen as taking place in a moment in time.

The chapter will elaborate scientific arguments for continuity or discontinuity of evolution, and what follows, for the existence or nonexistence of a clear borderline between our species and the rest of the living world. In turn, various possibilities of theological interpretations of the act of creation of man will be pointed out and a question will be considered to what extent theology is interested in a "momentary" account of this act.

After having cleared the respective positions of science and theology with regard to human origins, a particular proposal of reconciliation of the two views will be presented and its accuracy and acceptability will be reflected upon.

### Continuous Character of the Process of Evolution

The Darwinian theory of evolution—along, perhaps, with other non-Darwinian approaches to the process of evolution—aims at explaining the causes and mechanism of the process of change of biological organisms that occurs in sufficiently long periods of time and comprises many generations. Understood as such, it has little bearing on the evolution/creation controversy, and it may trigger discussions regarding only those causes and mechanism (like natural selection) that Darwin and his followers point at; specialists may look for other explanation of the leading factors of this process of change in the natural world that we witness around us. But this undoubtedly important matter may be a subject of consideration among evolutionary biologists and need not bother theologians or religious thinkers. The situation is quite different, however, when it comes to describing and interpreting the *course* of evolution, that is, phylogeny. Phylogenetic ancestor-descendant hypotheses aim at reconstructing the line of subsequent generations and show, for instance, how contemporary species are related to each other and to their common ancestors. The notion of a species plays an important role in those reconstructions—along the line of ancestors and their descendants, new species emerge from previous ones, and the latter either die out or develop further separately. This general scheme applies to all present or past species, and even if we are unable to follow a certain phylogenetic line (due, for instance, to poor fossil records), it is assumed that such a line existed. This applies obviously also to the species of *Homo sapiens*. A question of human origins posed on biological ground is therefore the question of the phylogenetic line leading to the species of *Homo sapiens*. Given this line, biology may also ask when and where there occurred such a decisive change in a population of our ancestors that resulted in our *really new* species. The problem is, however, that there seems to be no decisive criteria for distinguishing one species from another and therefore the very notion of *really new* one is, to some extent, empty. The conventional character of the notion of species may be shown on many levels. We will consider paleontology, genetics, and anthropology as disciplines where this is particularly obvious.

### The Notion of Species in Paleontology

Whereas in many other branches of biology the notion of species may be defined in an objective way (like the "classical" reproductive definition: two organisms belong to the same species if they can have fertile progeny), such definitions are inapplicable with regard to fossil populations. To use the notion of species in paleontology in a meaningful way, a statistical analysis has to be applied. Among organisms that reproduce sexually,[1] the distribution of most characteristic morphological features in a population is close to normal. Such features change gradually in subsequent populations, but when analysis shows that a certain new feature appears (or another ceases to be present) in a statistically significant part of population A, in comparison to population B (usually within a standard diversion [dispersion] in Gaussian distribution, which comprises 68 percent of population representatives), we may classify organisms belonging to A as representatives of different species than those belonging to B. This method is very useful, but a "borderline" that we may draw between the two species using this method is clearly conventional—a discrete notional framework is imposed on a continuous picture of gradual changes.

The significance of this paleontological way of distinguishing between different species for the problem of human origins stems from two facts. When looking down our phylogenetic line, we search for a "moment" or a "place" in the history of life on our planet when or where the first human population appeared. Hoping for an answer, we have no other option than to resort to paleontological methods and the notion of species embedded in them. What follows, what we may hope to find, is not a kind of "*true*" beginnings of humankind, but the beginnings of a species that we previously *decided* to define in a chosen way. What is more, the conventional character of the notion of species is not just a choice that we make for practical reasons, but it is a consequence of the continuous character of the process of evolution. The observations of evolutionary changes in fossil populations imply that there are no significant differences in the course of evolution—it displays continuity regardless of the environment where evolutionary processes were at work, or of the degree of development of organisms subject to these processes—which applies also to primates, including man (Dzik 2007). It seems therefore, that we are not only unable to point to an objective moment when the first humans appeared, but that such a "moment" never happened.

### Genetic Borderlines between Species?

A reconstruction of our phylogenetic past with the use of methods of molecular biology is directly impossible. Human ancestors died out and what is at our disposal now is only fossil remains from which we can obtain only fragments of mitochondrial DNA. What can be done, however, is an analysis of a genome sequence of contemporary species. Such analysis can show the genetic basis of the differences between species and, with the use of additional, usually noncontroversial, assumptions can suggest possible evolutionary changes. The human genome was sequenced a few years ago and the chimpanzee genome sequence was published in 2005. A comparison of these two genomes shows about five hundred genes in which the DNA differences are the biggest and therefore probably decisive for morphological, behavioral, and other differences between the two species (Stępień 2007). Although we are not chimpanzees' descendants, but have with them a common ancestor, a comparison of our genomes may suggest possible evolutionary changes that led eventually to our two so different species. We may assume that the differences between humans and their direct ancestor (regardless of the fact that we cannot be sure at present what particular species it was) were of similar kind and span. If we wanted to find out what particular change in molecular structure of our ancestor's DNA yielded a new species—*Homo sapiens* in this case—it wouldn't be plausible to attach this "moment" of a new species' arrival to a change in one particular gene. If, in turn, we liked to think that it was a more substantial part of the genome, we would have to decide where to draw the ultimate borderline: at the level of 10, 100, 300, or any other number of those genes in which we differ from our supposed ancestor. Quite clearly, a borderline we look for is again conventional—it is impossible to *find* it, but instead, we have to *decide* where we want to draw it.

## Anthropology and Its Methods

A look at some methodological issues of anthropology and examples of anthropological classification of certain discoveries shows that even this discipline, which deals directly with the past and origin of *Homo sapiens*, does not have—and perhaps cannot have in principle—any "proper definition" of man. And if this observation is correct, any answer to the question of a place and time when our species came into existence can be, again, only conventional.

There are three basic ways of approaching the past of our species, approaches that are used in three anthropological methods: the analysis of fossil remains (the morphological method), the interpretation of artifacts (the archeological method), and genetic analysis (the genetic method).

The first of these, and the oldest one used in anthropology, is the morphological method that tries to determine the degree of relationship between a fossil organism and a contemporary one by a comparative analysis of bone material. There are various morphological criteria that are used to classify fossil remains as belonging to a representative of a particular species. What is taken into account when classifying certain remains as representing the ancestors of *Homo* is, for example, the morphology of dental structures, the phenomenon of bipedality (which is inferred from the structure of certain bones), or the volume of the braincase. Using this method, the first representatives of our species (so-called early archaic *Homo sapiens*) were identified in some remains dated up to four hundred thousand years in age (Stringer et al. 1984; Aiello 1995).

The archeological method searches for the degree of phylogenetic similarity on the basis of the behavioral pattern recorded in such forms of material evidence as tools, burial places, objects of arts and crafts, etc. (Tomczyk 2006). To infer the behavior from such forms of material evidence as these, and to interpret this behavior as having specified meaning for a population that probably displayed it, one has to adopt quite strong auxiliary hypotheses, and the epistemological value of such reasoning depends on the accuracy of those hypotheses. This problem is, however, an internal issue of this particular method and justification of its credibility has to be left to specialists using this method in their anthropological consideration. From the point of view of this chapter, we need only to note that this method leads to the conclusion that the history of humankind began only thirty to forty-five thousand years ago based on evidence of human existence such as cave paintings and quite sophisticated tools (Clark 1967; White 1989; Harrold 1992; Leakey 1995).

The last of the anthropological methods mentioned is the genetic one. It "defines genealogy, projecting the degree of biochemical-serological as well as genetic affinity on the temporal axis" and tries to answer such questions as: "Which functional qualities of proteins, what kind of transformation of proteins, and what changes in the sequences of the nucleotides stand behind the origin of man?" (Tomczyk 2006). Analyses undertaken with the use of this method push the origin of our distant ancestors (the separation of human evolutionary branch from the primate tree) as far back in time as six to seven million years ago (Tattersall 2001; Tobias 2003; Lewin and Foley 2004), or even more (Tempelton 1993), and points at the molecular unity of Modern Man as established about two hundred thousand years ago (Cann et al. 1987).

Anthropological controversies that are triggered by the very fact of using different methods may be well illustrated by the history of solutions concerning the taxonomic status of the Neanderthal man. The morphological method seemed to show that although it belonged to the genus *Homo* (big braincase), it was not a representative of *Homo sapiens* (massive skeleton) (e.g., Keith 1931; Howells 1945). This opinion was called into doubt, however, when—with the use of the archeological method—the Neanderthal man had to be classified as an extinct variety of contemporary man when it was shown that the Neanderthals buried their dead and even had certain burial rituals. According to many scientists, such rituals prove that those who display them had to possess an essentially human notion of transcendence (e.g., Solecki 1971; Trinkaus and Shipman 1993; Stringer and Gamble 1993).[2] The result of these discoveries and its interpretations was that the Neanderthal man was no longer seen as a species different from *Homo sapiens* (*Homo neander-*

*thalensis*), but as a subspecies of our own (*Homo sapiens neanderthalensis*).

The dependence of the "definition" of "true man" on the assumptions applied in the respective anthropological methods can be seen not only "across" those methods, but also within them. An example of this is a short-lived carrer of *Oreopithecus* as the ancestor of contemporary humans (Tomczyk 2006). When the morphological method was applied, this creature was included in the family of *Hominidae*, because of its dental characteristic (Hürzeler 1960; 1968; Schaefer 1960; Straus 1963). This classification was, however, rejected (Tuttle 1975) when within the same morphological method, the ability to sustain upright position began to be demanded for any creature pretending to be a hominid form.

The diversity of research methods in anthropology is not surprising, nor is the fact that the "definition" of man is method-dependent. Even if some degree of consensus between the advocates of those methods can be achieved and a univocal definition of "true man" formulated,[3] it necessarily would be a conventional one. The choice of criteria (be they morphological, archeological, genetic, or other) according to which certain past forms are classified as our ancestors or their status of being early representatives of our species is denied, will be always arbitrary. And this is not a weakness of scientific methods, but a consequence of applying a discrete framework to describe an apparently continuous evolutionary process that eventually brought our species to existence. Our phylogenetic line is smooth—there are no objective borderlines or breaking points to which we could attribute "the moment and place" of the origin of man.

## Animals and Man: Borderlines?

Although many biological sciences suggest clearly, as it was demonstrated above, that there are no sharp differences between *Homo sapiens* and the rest of the living world, it can also be shown that humans do differ from their closest animal cousins. Contemporary knowledge of various behavioral sciences, like ethology or comparative and evolutionary psychology, allowed J. A. Chmurzyński to suggest a hypothesis of "continuity marked with leaps" (Chmurzyński 2007). There are many behavioral features that we share with the animal world. For example, many emotional homologies[4] can be found between man and apes—both in their causes and expressions. One of them is displacement activities, when in a conflict situation some people tend to scratch their heads or tidy their hair with fingers (Tinbergen 1977; Eibl-Eibesfeldt 1975). Even such forms of human activity that can be called "cultural" can be found in animals. Chimpanzees, for example, learn how to use certain tools from other representatives of their local population (Whiten et al. 1999; Whiten and Boesh 2001)—in a quite similar way as humans learn many forms of behavior from their social group—not as a part of their genetic inheritance, but by tradition. Even an ability to lie was found during experiments on sign language communication with a gorilla (Patterson 1978) and prostitution (in exchange for food) among bonobo chimpanzees (Cramb 2007).

All such homologies can be seen as yet another proof of the nonexceptionality of *Homo sapiens*. On the other hand, however, there are also such forms of human behavior that seem to have no prehuman precedents. For example, although animals can be furious and violent toward their enemy, it is a specifically human "ability" to experience persistent hatred. Only humans know what is shame, exercise trade (Grzegorczyk 1983), have syntactic speech, art, technology, and agriculture (Diamond 1992), and are inclined to transcendence, magic, or religion and search for generalized worldview (Wierciński 1994). What is more, there are also such typically human features that are in opposition to our etho-psychological inheritance. Some principles of human religious or ethical systems are in accordance with the biological principle of fitness maximization that gave rise to well-known explanations of human morality that attribute it exclusively to biological, evolutionary factors. But there are also such moral principles that not only do not give any evolutionary advantage, but, on the contrary, are clearly in opposition to biological needs of a species: the prohibition against stealing or lying, the condemnation of nepotism or (male) promiscuity, and the call to behave in a way that is unprofitable for an individual or his relatives: to tell the truth regardless of the circumstances, to keep and fulfill promises, to be altruistic toward nonrelatives without any reward (Bielicki 1990; 1993).

Seeking for essential differences between the animal and the human, many authors also point to our intellect as something that we do not share with the rest of the living world. This is more controversial, since some mental abilities seem not to be exclusively human. It certainly requires some form of thinking to prepare tools with the use of material "consciously" searched for, which has been observed in some ape species (Chmurzyński 2002). Even abstract thinking was found in the animal world—chimpanzees can be taught to count and express the result in figures, including the use of "zero" (Boysen and Berntson 1989).

Also, the argument that it is an exclusively human ability to recognize necessary truths—mathematical and logical (e.g., Barr 2003)—can be called into doubt. It is indeed hard to imagine that this kind of cognition could be possible even among the most "intelligent" apes, but the very notion of necessary truths can be questioned. If mathematics and logic is just a human construct, to some extent nothing more than syntactic "play" with conventional rules (despite contrary, neo-platonic views, such a position can be defended), then there is nothing special in "recognizing" that, for example, 1 + 1 = 2. This "truth" is in fact a thesis that can be proved from a set of arbitrary definitions and axioms. It is not convincing to make the case that when a human child does such a calculation, s/he *understands* what it really means, whereas "counting chimpanzees" can only *learn* how to do such calculations but do not understand them (Barr 2003). In fact, humans too learn how to use certain symbols, and there is nothing "obviously understandable" in the above formula. Only by training do we "understand" that to have one chocolate bar and then another chocolate bar is the same as to have two chocolate bars. We suspect that for every glutton there is a huge difference between having both treats at once and having them one by one.[5]

On the other hand, our other mental abilities do seem to give us an exceptional position in the animal world. People use abstract notions like "femininity" or "circularity" and understand them as referring to "ideas" as opposed to concretes (Barr 2003).[6] Also, a characteristic feature of human language, which K. R. Popper calls its argumentative function (enabling us to confirm or falsify previously formulated theses [Popper 1979; cf. Przechowski 2007]) may be seen as something distinctively human (though the other Popperian highest function of language—description—can be attributed to some forms of animal communication).

Many authors try to explain human thinking in terms of the evolutionary advantage of *Homo sapiens*, and thus maintain that although our mental abilities are exceptional, they do not suggest any essential "gap" between us and animals. Popper, for example, speaks of the evolutionary origin of his highest functions of language. But at least one human mental feature seems difficult to explain in this way. This is an ability—or at least, inclination—to pose and answer "purely theoretical" questions—to solve problems, which solutions have no practical consequences. We want "to know in order to know," not only "to know in order to use." Aristotle thought that such "pure" knowledge is much more desirable than practical knowledge, because only it is worthy of man.

### "Punctual" Creation?

For many nontheologians, creation is a unique, supernatural act of God,[7] taking place a very long time ago when the Creator brought certain beings—or the world itself—into existence. This view of a special moment in time when the divine act is performed is subject to challenge not only by current theories in cosmology, but also by long-standing theological opinions, first formulated in ancient Christianity by St. Augustine, that God created the world not *in* time but *together with* time. But with regard to the creation of man—as a basic reading of the book of Genesis seems to suggest—after the creation of the world and the rest of its beings, the problem of the nonexistence of time at "the moment" of creation disappears, and the creation taking place at some moment in time is back on the agenda.

Such a view of creation of man was one of the reasons why the Darwinian theory of evolution was at first quite strongly opposed by many Christian thinkers.[8] What seemed especially difficult to accept was the alleged consequences of the evolutionary view of human origins for the Christian doctrine of the original sin. When it became clear that from a biological point of view it is impossible to hold that at the beginning of our species there was numerically one pair of "the parents of everybody," the doctrine of the original sin, which seemed to require monogenism, had to be challenged. Certain solutions were found, however, especially after the encyclical *Humani generis* issued in 1950 by Pope Pius XII.[9] Although the pope pointed out that polygenism seems to contradict certain elements of Christian doctrine, at the same time he allowed that proper investigation can be carried out in all disciplines, including theology, with regard to the evolutionary origin of the human body. This opened a path to new understanding of the dogmatic truth of original sin. If it can be accepted that humanization occurred by evolution, then both "first parents" originated in the same way and polygenism cannot be avoided at least with regard to the first human pair. In turn, if two human beings came independently from the animal world, there is no reason why it must have been only two of them and not more (Anderwald 2007). All humans were originally in the state of biological-historical unity, and therefore it is possible that one human (or one pair) committed the sin and, because of that, the rest of united humanity was deprived of its holy state of God's grace. Alternately, the whole of humanity—of polygenetic origins—committed the sin in the persons of all its members as a group, historically one and united (Rahner 1967). As it also became clear, in the evolutionary perspective it is impossible to define a historical time and place when the original sin was committed (Schmitz-Moormann 1969).[10] And because original sin is an intrinsic part of the whole doctrine of the creation of man, it is equally unconvincing that speaking theologically about the origins of man we have to search for *the moment* and *the time* when the act of creation of man took place.

The problem with the doctrine of original sin brought about by biological rejection of the notion of monogenism was eventually considered from a broader perspective and theology returned to concepts that have been known in fact from the beginnings of Christianity.

One of the key principles of Christian theology since its beginning was a differentiation between the truths of the faith and the form of their presentation. As early as the second century, St. Irenaeus, commenting on the "story" of creation of man by "shaping him from the soil of the ground" (Gen. 2:7), wrote that God "shaped man with his own hands, that is through Son and the Holy Spirit" (quotation after Salij 2007). Ancient Christianity knew that the Bible presents important truths in an anthropomorphic way, and it would be naïve not to separate the meaning of the scriptural teaching from its anthropomorphic form. The symbol of "the soil of the ground" from which man was shaped was understood as showing that man is a part of nature. In the Middle Ages, such an approach to the biblical "stories" was further developed in St. Thomas Aquinas' theology of creation. Aquinas taught that God gives his creatures a share in his own causal power—being the immediate cause of the whole world and every individual being, he allows some creatures to be causes of other ones (Salij 1995). Also the problem of the time span of the created world was of secondary importance for Aquinas. Although he personally believed that the world had its beginning in time, it seemed equally possible to him that it existed eternally. The fundamental meaning of the belief in creation is not that the world came into existence at a particular moment in time, but its continual relation to the Creator. And this relation could last eternally (*Summa Theologica*, Part I Question, 46).

It is indeed surprising, that although theology had had all those sophisticated tools at its disposal for centuries before Darwin, they all were forgotten when the controversy between the evolutionary explanation of the development of the living world and a religious belief in creation emerged in the nineteenth century. Most Christian thinkers in the time of Darwin tried to defend a common view of the Creator who created all beings by means of giving existence to the first representatives of every single species. Under the pressure of the theory of evolution, some theologians replaced this view with the doctrine of special divine interventions in the crucial moments in the evolutionary development of the world. While for the majority of its history, the world could—according to this doctrine—be governed by natural forces driving its evolutionary course, at least two moments had to be exempt from the rule of evolution: the origin of life and the appearance of the first human. Such a view, although more advanced than a simple picture of God, the craftsman who builds its creation step-by-step, was not only insufficient from an evolutionary point of view, but also theologically inadequate. The need for special interventions by the Creator in the course of the history of the world may suggest a false view of God who is external to the world. Whereas God, while being transcendent, is also present everywhere and in every moment: "In him we live and move and have our being" (Acts 17:28) (Salij 2007).

Many of the bitter disputes at the end of the nineteenth and at the beginning of the twentieth century could have been avoided had theologians remembered that the dogma of creation does not require a moment in time when creation (of the world or of particular species, including humans) takes place. It holds, instead, that everything that exists is continually given its existence by God. A timely beginning of the world or of man is irrelevant. The world and man were created, which means that creatures are dependent in every moment of their existence on God. The world is able to develop driven by natural forces and to discover and describe those forces is a task of natural sciences. It is being given its existence *as such* (able to develop) by God, and this is the subject of faith in creation (Salij 2007).

### A Need for a Solution?

As was demonstrated above, one of the main issues in the evolution/creation controversy, a discordance between the continuous evolutionary approach and the "punctual" religious (theological) view of creation, is not so sharp as it might seem. Neither do scientific data force us to admit that because *Homo sapiens* is a product of evolution, our species does not differ in an essential way from the rest of the living world, nor does the theological account of creation require a "punctual" understanding of this divine act. A solution—if there is still need for it—should be searched for beyond this not-so-sharp opposition between "continuous evolution" and "punctual creation."

Since there have been several attempts to look at evolution and creation in a unifying manner, it is worthwhile to consider them from our "weakened opposition" perspective and see if those solutions can be accepted in light of our interpretation of scientific data, on the one hand, and a proper theological understanding of creation on the other. An example of such attempts is the "evolutionary model of creation" developed by Polish philosopher K. Kloskowski (1994).

The model in question is based on two assumptions: (1) a process interpretation of the world and, especially, the living world, and (2) an epistemological choice of the evolutionary theory of knowledge, developed by R. Riedl (1981, 1984), which is seen as the best tool for describing and understanding reality as a process. Both assumptions are of a philosophical kind, which shows that a solution to our controversy can be reached outside of purely scientific or purely theological perspectives. Those two realms of knowledge rightly enjoy methodological and epistemological independence, hence any kind of a "unifying view" of particular knowledge or concepts formulated in both of them requires a "third party" providing a ground and tools for such "unification." In our case, this "third party" is philosophy. We need to accept certain philosophical presumptions, and what we can eventually obtain is not a changed scientific or theological concept, but a philosophical worldview based on those two.

Assumption (1) is a choice that implies a particular ontological perspective in which fundamental ontological entities are not things or events, but processes—the world itself is a process; it *is* not, but constantly *becomes*. Such a view allows for both an evolutionary, continuous view of the living world (which is clearly suggested by most biological sciences), and a theological account of creation that does not stress a moment in time when certain beings came to existence but understands the truth of creation as a conviction of continuous dependence of creatures on their Creator.

Assumption (2) serves mainly to show the essential differences between humanity and the rest of the world and hence to justify the need for a special act of creation of man. Evolutionary epistemology maintains that our cognitive abilities evolved under the pressure of natural selection. They appeared due to the natural influences of the world on our ancestors and we enjoy them because they serve our evolutionary success in this world. This implies that our cognitive abilities are "in accordance" with the world (if they were not, they would not have been chosen for by evolution). This means that generally our cognition has to be correct, because there is a sort of isomorphism between the pattern of nature and the pattern of our thought and cognition (Kloskowski 1994). In turn, since apart from natural cognition, humans also developed spiritual cognition, the latter has its sources (according to the theses of evolutionary epistemology) in reality. Asking about the origin of man, we have to look for an answer in both natural and spiritual reality, because both "realities" are mirrored in human cognitive capacities and an answer based only on one of them would be inadequate.

The two assumptions described above allow Kloskowski to suggest that evolution can be seen as a specific "moment" of the act of creation. An act of the creation of man is required to account for the spiritual side of human reality mirrored in our cognition. But because our spirituality appeared in the course of evolution, and, according to assumption (1), the whole world, including ourselves, is a continuous process, we cannot find a "moment" when the act of creation happened. Instead, evolution must be seen as a process occurring *within* the act of creation and may be called a "moment" (meaning part) of this act.

It seems that Kloskowski's proposal does indeed go beyond the simple opposition, challenged in this chapter, between an absolutely continuous account of evolution and a momentarily understood act of creation. But the assumptions that allow him to put forward his final thesis of evolution as a "moment" of creation are debatable. Obviously, if we agree that the solution of the evolution/creation controversy has to be of philosophical character, we have to define the philosophical basics of the proposed approach. Such basics are subject to many choices and it is impossible to evaluate them in a fully objective manner. However, it would be interesting to see if Kloskowski's assumptions could be weakened without denying his conclusions. Particularly, applying the consequences of the controversial evolutionary epistemology seems both insufficient and unnecessary. The very existence of human spiritual cognition can be challenged. Moreover, perhaps there is no need of any "proof" of the special spiritual abilities of *Homo sapiens* and what is sufficient for our purpose is to note essential differences (relying not necessarily on our spirituality) between humans and animals. And some exceptional characteristics of humanity do seem to exist, which was shown earlier.

As for the application of process ontology, the question is whether a similar construction can be achieved without those particular ontological choices. It is true that one of the fundamental features of reality is its changeability. But one may want to maintain that what changes are *things*, and in such an ontological perspective we also should be able to demonstrate the possibility of concordance between a not-so-continuous evolutionary view of the origins of *Homo sapiens* and a not-so-punctual account of the creation of man.

## Conclusion

This chapter drew upon one particular problem in the evolution/creation controversy—the tension between the continuous character of evolutionary processes and a punctual understanding of the act of creation. It was demonstrated that this opposition is not so sharp as it may seem. So any search for a solution to our problem has to go beyond this opposition and not try to reconcile continuous evolution (because it is not absolutely continuous) with punctual creation (because it does not need to be—or, indeed, cannot be—understood, for theological reasons, in such a restrictive way).

Apart from the particular issue that we have been concerned with in this chapter, there are also others that appear in the details of the allegedly conflicting theological and scientific views on the origin of man. Other important issues include the problem (mentioned in the chapter) of original sin versus polygenism and the problem of chance as a driving force of evolution versus a causal and final character of creation. Although these problems can and ought to be distinguished, they are interrelated. Hence, a good proposal for the reconciliation of theological and scientific perspectives on the origin of man should offer a tool for possible solutions to all of them.

## Notes

1. Organisms that reproduce in a nonsexual way are classified in an even more conventional way.
2. In such a pattern of reasoning—from material remains of burial places and those suggesting burial rituals to granting the possession of "essentially human notion of transcendence"—we can see that certain auxiliary hypotheses mentioned above have to be used in interpretations of what is excavated.
3. Whether it is possible or not is irrelevant for the main conclusion of this paragraph. We do not think either that such univocal definition should be required.
4. Behavioral homologies are such features that can be found in all related species and are similar with regard to their form and origin, though they may differ in their functions (Meissner 1976). The example given above shows such a form of animal and human behavior, which is homologous and similar also in its function.
5. Obviously there are much more serious arguments against the idea of necessary truths, but these are well known in the history of philosophy.
6. A position one takes in the medieval controversy over universal notions is irrelevant here. We use and understand such notions regardless of what we think they actually refer to.
7. Preliminary results of recent research of opinions among Polish students and teachers of biology showed (the final results are yet to be published) that the majority of them hold such a view.
8. It was not the only reason of this opposition, though. A much more important reason for the Christian reluctance toward Darwin's theory was a fear that it undermines human exceptional position among the rest of the creation and man's dignity stemming from his likeness to the Creator (Gen. 1:26–27). It is also worth noting that none of Darwin's books were ever put on the Index, although in that time many books were all too easily regarded by the church as unacceptable. Hence the opinion that the church in its official decisions was in strong opposition to Darwinism is more an artifact made up by antitheist writers than a true report on facts (Salij 2007).
9. We are speaking here about proposals formulated in Catholic theology.
10. Despite those attempts to reinterpret the doctrine regarding original sin, official teaching of the Catholic Church admits that the issue of the "transmission" of original sin from the parents of humanity to all its members remains a mystery (Catechism of the Catholic Church). It seems that theological research of this problem is still required.

## Bibliography


Aiello, L. C. "The Fossil Evidence for Modern Human Origins in Africa: A Revised View." *American Anthropologist* 95 (1995): 73–96.

Anderwald, A. "Początki człowieka a grzech pierworodny. Od konfliktu do integracji" ["The Beginning of Man and the Original Sin"]. In *Kontrowersje wokół początków człowieka* [*Controversies about Human Origins*], edited by G. Bugajak and J. Tomczyk, 287–97 (Katowice, Poland: Księgarnia św. Jacka, 2007).

Barr, S. M. *Modern Physics and Ancient Faith.* Notre Dame, IN: University of Notre Dame Press, 2003.

Bielicki, T. "O pewnej osobliwości człowieka jako gatunku. [On Certain Peculiarity of Man as a Species]" *Kosmos* 39, no. 1 (1990): 129–46.

———. "O pewnej osobliwości człowieka jako gatunku [On Certain Peculiarity of Man as a Species]." *Znak* 45, no. 1 (1993): 22–40.

Boysen, S. T., and G. G. Berntson. "Numerical Competence in a Chimpanzee (Pan troglodytes)." *Journal of Comparative Psychology* 103 (1989): 23–31.

Cann, R. L., M. Stoneking, and A. C. Wilson. "Mitochondrial DNA and Human Evolution." *Nature* 325 (1987): 31–36.

Chmurzyński, J.A. "Etopsychiczne granice między zwierzętami a człowiekiem" ["Behavioral and Mental Borders between Animals and Man"]. In *Kontrowersje wokół początków człowieka* [*Controversies about Human Origins*], 27–42.

———. *Szczeble zdolności poznawczych w świecie zwierząt. Rozważania behawioralne i zoopsychologiczne* [*Grades of Cognitive Capacities in the Animal World*]. Warsaw, Poland: Instytut Biologii Doświadczalnej im. M. Nenckiego PAN, 2002.

Clark, G. *The Stone Age Hunters.* London: Thames & Hudson, 1967.

Cramb, A. Female Chimpanzees "Sell" Sex For Fruit. http://www.freerepublic.com/focus/f-chat/1896710/posts. Accessed November 9, 2007.

Diamond, J. M. *The Third Chimpanzee: The Evolution and Future of the Human Animal.* New York: HarperCollins, 1992.

Dzik, J. "Sposoby odczytywania kopalnego zapisu ewolucji" ["Methods of Reading the Fossil Record of Evolution"], *Kontrowersje wokół początków człowieka* [*Controversies about Human Origins*], 65–86.

Eibl–Eibesfeldt, I. *Ethology: The Biology of Behavior.* 2nd ed. New York: Rinehart and Winston, 1975.

Grzegorczyk, A. "Antropologiczna wizja kondycji ludzkiej." ["Anthropological Vision of Human Condition."] *Roczniki Filozoficzne* 31, no. 3 (1983): 59–81.

Harrold, F. B. "Paleolithic Archeology, Ancient Behavior, and the Transition to Modern Homo." In *Continuity or Replacement Controversies in Homo sapiens Evolution*, edited by G. Bräuer and F. H. Smith, 219–30. Rotterdam: A. A. Balkoma, 1992.

Howells, W. W. *Mankind So Far.* New York: Doubleday, Doran & Company, 1945.

Hürzeler, J. "Questions et Réflexions Sur L'Histoire des Anthropomorphes." *Annales de Paléontologie* 54 (1968): 195–233.

———. "Signification De L'Oréopithèque Dans La Phylogénie Humaine." *Triangle* 4 (1960): 164–74.

Keith, A. *New Discoveries Relating to the Antiquity of Man*. London: Williams & Norgate, 1931.

Kloskowski, K. *Między ewolucją a kreacją* [*Between Evolution and Creation*]. Warsaw, Poland: ATK, 1994.

Leakey, R. *The Origin of Humankind*. London: Phoenix, 1995.

Lewin, R., and R. A. Foley. *Principles of Human Evolution*. Oxford: Blackwell, 2004.

Meissner, K. *Homologieforschung in der Ethologie*. Jena, Germany: G. Fischer Verlag, 1976.

Patterson, F. "Conversations with a Gorilla." *National Geographic* 154, no. 4 (1978): 438–65.

Popper, K. R. *Objective Knowledge: An Evolutionary Approach*. Rev. ed. New York: Oxford University Press, 1979.

Przechowski, M. "Zagadnienie ewolucji w ujęciu K. R. Poppera" ["Issue of Evolution according to K. R. Popper"]. In *Kontrowersje wokół początków człowieka* [*Controversies about Human Origins*], 163–73.

Rahner, K. "Erbsünde und Evolution." *Concilium* 3 (1967): 459–65.

Riedl, R. *Biologie der Erkenntnis*. Berlin/Hamburg: Paul Parey, 1981.

———. *Die Strategie der Genesis*. München/Zürich: Piper, 1984.

Salij, J. *Eseje tomistyczne* [*Tomistic Essays*]. Poznan, Poland: W drodze, 1995.

———. "Pochodzenie człowieka w świetle wiary i nauki" ["The Origin of Man in the Light of Faith and Science"]. In *Kontrowersje wokół początków człowieka* [*Controversies about Human Origins*], 277–86.

Schaefer, H. *Der Mensch in Raum und Zeit mit besonderer Berücksichtigung des Oreopithecus-Problems*. Basel, Switzerland: Naturhistorisches Museum, 1960.

Schmitz-Moormann, K. *Die Erbsünde. Überholte Vorstellung und bleibender Glaube*. Freiburg im Br., Germany: Walter-Verlag, 1969.

Solecki, R. S. *Shanidar—The First Flower People*. New York: A. Knopf, 1971.

Stępień, P. P. "Ciągłość czy moment—rozważania genetyka" ["Moment or Continuum? A Geneticist View"]. In *Kontrowersje wokół początków człowieka* [*Controversies about Human Origins*], 23–26.

Straus, W. L. "The Classification of Oreopithecus." In *Classification and Human Evolution*, edited by S. L. Washburn, 146–77. Chicago: Aldine, 1963.

Stringer, C. B., and C. Gamble. *In Search of the Neanderthals*. London: Thames & Hudson, 1993.

Stringer, C. B., J. J. Hublin, and B. Vandermeersch. "The Origins of Anatomically Modern Humans in Western Europe." In *The Origins of Modern Humans: A World Survey of the Fossil Evidence*, edited by F. H. Smith and F. Spencer, 51–135. New York: Alan R. Liss, 1984.

Tattersall, I. *The Human Odyssey: Four Million Years of Human Evolution*. New York: Universe.Inc., 2001.

Tempelton, A. R. “The ‘Eve’ Hypothesis: A Genetic Critique and Reanalysis.” *American Anthropologist* 95 (1993): 51–72.

Tinbergen, N. *Study of Instinct*. Norwood, PA: Norwood Editions, 1977.

Tobias, P. V. “Twenty Questions about Human Evolution.” Human Evolution Conference Proceedings, XV-ICAES Florence, Italy, 2003: 9–64.

Tomczyk, J. “The Origin of ‘Homo Sapiens’ in the Light of Different Research Methods.” *Human Evolution* 21 (2006): 203–13.

Trinkaus, E., and P. Shipman. *The Neanderthals: Changing the Image of Mankind*. New York: A. Knopf, 1993.

Tuttle, R. H. *Paleoanthropology: Morphology and Paleoecology*. The Hague, Paris: Mouton Publishers, 1975.

White, R. “Visual Thinking in the Ice Age.” *Scientific American* 7 (1989): 92–99.

Whiten, A., and C. Boesch. “The Cultures of Chimpanzees.” *Scientific American* 284 (2001): 48–55.

Whiten, A., J. Goodall, W. C. McGrew, T. Nishida, V. Reynolds, Y. Sugiyama, C. E. G. Tuzin, R. W. Wrangham, and C. Bosch. “Cultures in Chimpanzees.” *Nature* 399 (1999): 682–85.

Wierciński, A. *Magia i religia. Szkice z antropologii kultury* [*Magic and Religion*]. Cracow, Poland: Nomos, 1994.

Editor's Introduction to

# Mathematics as a Formal Ontology: The Hermeneutical Dimensions of Natural Science and Eastern Patristics

**Alexei Chernyakov,** St. Petersburg School of Religion and Philosophy, Russia

Alexei Chernyakov builds the case for a hermeneutical understanding of scientific concepts. In continental thought and patristic doctrine, he finds the roots of an ontological understanding that allows analytical frameworks to shift and transform themselves as the dynamic process of creation continuously unfolds. Mathematics—too often misunderstood as a fixed *episteme*, reflective of unchanging Platonic forms—is, in fact, constantly recreated. It is a "formal hermeneutic" whose designations evolve in reference to the ongoing refinements and discoveries of the "facts" of nature.

Chernyakov reminds us that the "first philosophy" is an inquiry into being *qua* being, an inquiry that is, by its very nature, about the Divine. So it is not at all surprising to find theological and philosophical approaches productive in investigating the mathematics that underlies the modern scientific project. The same openness to the permanent change of the human experience that is found in the constant refinement of mathematics resonates there. In fact, biblical hermeneutics and the patristic tradition make it clear that even "Holy Writ is an ambiguous concept . . . (that) does not exist autonomously without the tradition of exegesis."

The historical error of imagining that the facts of nature speak for themselves is no less erroneous for having been embraced by Galileo and Luther. But the hermeneutical approach, Chernyakov argues, helps us to put their assertiveness into context. Neither Luther's interpretation of *justitia Dei* nor Galileo's "Book of Nature" is a misguided interpretive base. But both exist as part of a hermeneutical trajectory that draws their ontological frames from a storied past into a changed future. And it is innovations in mathematics that carry that change forward, "the *modus existendi* of the scientifically interpreted world depends on the *modus existendi* of the objects of mathematics."

As with the morphing of the classical concept of *topos* into Newtonian space and Riemann's geometry, there is a constant historical replacement of formulae for the relationships of the world. The *nominata*, "the invariants of the historically changing context," pairs with the *nomena*, which "shifts in historical drifts from one context to another." Underlying this constant shifting, though, is the thread of a hermeneutic trajectory that replaces the one-dimensional Platonic being with an open, uncompleted process.

This hermeneutical vision was present in Eastern Christian theology from its earliest inception. The *logoi* were not simply energetic bits of divine presence but motive elements of a changeful universe, the "divine and good intentions for the world." Nature aspires to collaborate with God in fulfilling its own divine purpose, a collaboration that "is a part of the creative plan itself."

Science and mathematics reveal the hermeneutical ontology as they break through the tangible resistance to conceptual innovation. When discovery breaks, when unexpected facts emerge, these are the moments when the hermeneutical trajectory shifts and when the permanent processes of creation, the *logoi spermatikoi*, are at work.

Alexei Chernyakov is a member of the team working on "The Religious Basis of Contemporary Problems in the Natural Sciences and Humanities," a GPSS-award-winning project headed by Natalia Pecherskaya, rector of the St. Petersburg School of Religion and Philosophy.

P.D.

10

# Mathematics as a Formal Ontology

## The Hermeneutical Dimensions of Natural Sciences and Eastern Patristics

**Alexei Chernyakov,** St. Petersburg School of Religion and Philosophy, Russia

### Mathematics as a Formal Ontology

My immediate task is to characterize mathematics as a "formal ontology" and to demonstrate that there is a hermeneutical dimension to mathematics itself.

Modern natural sciences in their totality claim to be a universal ontology, though this claim lies perhaps outside the proper subject of science. Even if we reduce the ontological claim of science to the modest minimum, there is no doubt that science in accordance with the dominant contemporary world outlook considers itself as a universal ontology of nature. The huge progress made by the sciences since Galileo is, in many respects, a result of their new mathematical form. Mathematics becomes an inalienable part of the contemporary "natural philosophy." We could say that mathematics shapes it.

Extensive research has been devoted to the role of mathematics in the natural sciences. Our question is different. If mathematics is to be thought as *the* formal ontology of nature, then it seems to be natural to assume that the analysis of "the way of being" of mathematics itself, in particular the foundations of mathematics, should play the role of the most fundamental discipline that Aristotle called *the first philosophy* and his disciples, *metaphysics*.

(Let me mention in advance, that the first philosophy, according to Aristotle's explanation, asks about "being *qua* being" and can also be called "theology" because the question about the most basic principles of "what-is" is inseparable from the inquiry about the divine being. One has to look for the answer outside science in a variety of nuanced ways to address this kind of "ultimate question" to which philosophy and theology belong. At the end of this chapter I shall try to show how the idea of a "hermeneutical ontology" resonate with the patristic tradition of the Eastern Christianity.)

The hermeneutical approach allows us to comprehend two aspects of the contemporary *modus existendi* of mathematics within the corpus of science. The first concerns the choice of mathematical formalism as the skeleton of the modern scientific conceptual systems ("conceptual schemes"). The second is connected with internal structure of mathematics itself, and I shall discuss it in more detail later on.

The creators of "modern science" intentionally chose a "proper" language of natural philosophy. Now we may recall that this choice was motivated precisely by the idea that the universe is a kind of grand project realized by the divine Creator and that is developed in the language of mathematics. In one of his writings in 1623 Galileo stated:

> Philosophy [i.e., natural philosophy] is written in this grand book, the universe, which stands constantly open to our gaze. But the book cannot be understood unless one first learns to comprehend the language and read the letters in which it is composed. It is written in the language of mathematics, and its characters are triangles, circles and other geometric figures, without which it is humanly impossible to understand a single word of it, without these, one wanders about in a dark labyrinth.[1]

It is quite clear from this text that Galileo supposes the existence of a "literal sense" contained in natural phenomena and that he supposes there exists an "original language" that records it. Therefore the *only* "correct" language capable of describing the "laws of the universe" must be selected by scientists themselves. And, further, it implies that the *only* "correct" language allows for a scientific "description" of facts. If we want our description to be really and authentically "scientific," we must follow these rules.

But the tradition of biblical hermeneutics teaches us that the Holy Scripture remains in a permanent process of, let me say, "rewriting," which means that time and again it produces new essential meanings and values within the permanently changing human world and its varied cultural and intellectual contexts. This hermeneutical tradition teaches that the "authentic sense" of the Holy Writ is an ambiguous concept—if by that expression we refer to a "thing in itself," something akin to Wiggins' "horse, leaves, sun and stars" having an autonomous existence, independent of the traditions of interpretations. According to my tradition—to the attitude of the Orthodox Church—the "strict sense" of the scripture does not exist autonomously, without the tradition of exegesis.

The modern scientific search for the *only* "correct" language to express the sense (i.e., the "literal sense") of meaning in an absolutely univocal way is nothing less than an attempt at avoiding interpretation altogether. It is clear that this attempt was fostered by the inherent problems of the hermeneutical approach itself. This was already clearly manifest within the framework of Christian exegesis—for example, in the disputes of theologians at Alexandria and Antiochus. Thus certain questions must unavoidably be raised. To what extent can this or that interpretation (exegesis) be verified? What meaning can we *not* suppose (incredible) to be hidden behind the lines of Holy Writ? Is there a criterion for the universal relevance of our interpretation? Of course the church's tradition possesses certain long-standing modes of verification dependent on the *consensus patrum*. Yet these criteria rely on certain theological premises that can hardly be unanimously accepted across the breadth and width of all Christendom. Hermeneutics as a science sets itself precisely the goal of defending the text from the arbitrary ideas of this or that interpreter.

From the other side we would now like once more to draw attention to the impossibility of avoiding the hermeneutical dimension as such, that is, to its irreducibility. During the days when modern science was gaining converts to the language of mathematics, analysis of Holy Writ as dictated by church tradition became unsatisfying. And they said, "No—we shall address the text itself!" This was what Luther did. He rejected all interpreting texts and said, "No, I shall read the Holy Writ itself, *sola Scriptura*!" At a certain moment in history such a step may have been extremely fruitful and served to immensely widen the scope of our individual vision. Yet was it still possible that anyone could actually *read the Holy Writ itself*? May not this or that theological tradition represent *the very mode of existence* of the meaning of Holy Writ? Any attempt to discard all "other" interpretations does not necessarily lead to an understanding of the "literal sense" of the text; it just marks a transition to another hermeneutical tradition, which itself has not sprung up from scratch. For instance, Luther relied on the Latin version of St. Paul's Epistle to the Romans (1:17) and interpreted it as he did based on his understanding of the words *justitia Dei*—on the basis of his own purifying and "indisputable" experience of being "justified by faith."[2]

Galileo also said, "I shall discard the whole of Aristotle's physics, the whole of the tradition of describing the essence of natural phenomena and shall read instead the book of nature itself." Yet while he was relying for that reading on his own observations and experiences, which he believed to be indisputable (since, like Luther's faith, they seemed to him based on immediately convincing facts), Galileo only *guessed* what the real language of nature was and in how it speaks to insightful observers. Of course, nature had not remained "silent" before Galileo. Nature is always speaking with us and to us and in us. We write poems and novels about it. The Greeks were writing philosophical texts on nature in their time. But Galileo presumptuously said, "No—*I* know the original language of nature which is the language of mathematics." Yet in fact the choice of this language by Galileo was not without precedent either. He plainly relied on the above-mentioned notion that the universe was "written" by the Creator (in the language of mathematics). That he chose mathematics to express himself does not negate the theological underpinnings of his scientific approach. And just as in the case of theological exegesis, we can question the grounds on which this language was chosen, its limitations and the scope of its "explanatory power."

Though, as it has been said, we are not inclined to look for the only possible, "authentic" language with which to speak about nature, it is a matter of fact that mathematical language has become an indispensable constituent of modern science and this has had important consequences.

The choice of the language of a scientific theory inevitably means the choice of the language of factual description. So-called "facts" do not remain indifferent to the language in which they are described. I do not think that the notion of a "pure" fact, completely unconnected with the sphere of language, human activity, and human interests in general has any sense at all. In exactly the same way, the "literal meaning" of a text, without any interpretation, cannot be arrived at. The facts of science are loaded with theory and technology and a history of experiments. Natural sciences therefore do not escape the scope of hermeneutics any more than the humanities.

Galileo undertook the obligation to read the universe and its phenomena in a certain language that he considered to be *the authentic language* of the divine design, and contemporary science adheres to this project. It is exactly this idea that allows me to speak of mathematics as *the universal* formal ontology of modernity. But this "language" itself is by no means constant. To the contrary, it is subject to the impetus of historical change and conceptual innovation. This means that, on the one hand, natural phenomena, the subjects and the results of scientific observation, have acquired a privileged "conceptual scheme" with which to be comprehended and, on the other hand, that changes in the scientifically interpreted world depend first of all on conceptual changes in mathematics.

Sir Peter Strawson, while insisting on the existence of "a massive central core of human thinking which has no history," states at the same time that unchangeable concepts and categories "are not the specialties of the most refined thinking." On the contrary, "they are the commonplaces of the least refined thinking."[3] But mathematics does belong to the scope of the most refined thinking and must to all appearances pertain to the "specialist periphery" of vigorous conceptual change. It is exactly this zone that determines the conceptual schemes to "single out the things" of the world of sciences. If natural sciences in aggregate claim to be a universal ontology of nature, the *way of being* pertinent to the *beings* of this ontology depend on conceptual innovation in mathematics itself. In other words, the *modus existendi* of the scientifically interpreted world depends on the *modus existendi* of the objects of mathematics.

This discussion constitutes a massive part of the contemporary philosophy of mathematics, and the spectrum of different ontological positions here is rather wide. Hermeneutics, as it has been said, is both a new method and a new starting point in philosophy—a new "first philosophy" according to early Heidegger. The hermeneutical approach that, in the form of the "existential hermeneutics," is constitutive for Heidegger's fundamental ontology has its counterpart in the philosophy of mathematics. A detailed elaboration of this subject exceeds the limits of this work, but the main ideas can be outlined.

Recent attempts to overcome the canonical schism between Platonism, formalism, and constructivism,[4] and to understand mathematical entities within the framework of the empiricism of the good old days, to "modify the traditional account of [the objects of mathematics—sets, numbers, functions, etc.] as inaccessible Platonic things and instead bring them into our familiar space-time context," even to argue "that they are accessible to our ordinary perception,"[5] clearly contradict the hermeneutical ontology that proclaims a mutual dependency between "facts" and "concepts." Concepts, and mathematical concepts in particular, cannot be "read out" of the "facts." They cannot be a result of an "abstraction procedure" because, according to our explanation, they single out the "facts" and "cut up the [scientifically apprehended] world into objects." On the other hand, it is likewise impossible to ascribe to mathematical entities a kind of eternal, ahistorical, self-sufficient being—that is, to adhere to a Platonic ontological position. There is a hermeneutic alternative to Platonism in mathematics that is quite different from the neoempiricism mentioned above. According to this philosophical position, *mathematics itself is to be understood as a formal hermeneutic.*[6] There is a set of *names* "designating" from time immemorial the privileged subjects of mathematical thought, such as "number," "infinity," "space," "continuum," etc. What are the *nominata* for these *nomena*, and what is their way of being? Of course, they do not exist in space and time as objects of sense perception. And, as it has been mentioned, I do not think that in relation to them an "empirical reduction" is possible. Let me repeat the ontological formula that has been phrased above: the *nominatum* of a "mathematical *nomen*" is nothing other than the invariant of the historically changing contexts of the *nomen*-usage. These manifold contexts are bound together by the relations of (sometimes mutual) interpretations, that is, by "hermeneutical relations." The *being* of a *nominatum* is nothing else than the identity of its "hermeneutical trajectory" within history, within variegated, historically changeable conceptual schemes. A *nomen* shifts in its historical drift from one context to another. Within a certain context it has a more or less definite meaning, but this meaning can be drastically changed in the course of this conceptual and contextual journey. However, a certain succession is preserved, a certain continuity of a "hermeneutical trajectory" is sustained, which can be disclosed by appropriate research into the history of ideas. Let us consider for example the concept of *topos* or *chora* in Greek mathematics, the notion of *space* in Newtonian mechanics (and its purely mathematical counterpart—*the* Euclidean space), and then the chain of their "descendant" concepts, such as, for example, the Riemann manifold, "scheme" (in the sense of the contemporary algebraic geometry), Grothendieck *topoi*, etc.

The relation between the previous and subsequent concept, between the ancestor and the descendant is not that of "generalization." It has a much more complicated, though in each case clearly determined, character. In many cases, for example, such a relation can be grasped formally as a "functor" between two different categories (in the sense of Mac Lane's *category theory*). Even within a synchronic layer of mathematics, there is a system of different formal approaches, different axiomatic systems, etc. These form a set of perspectives to look at an indeterminate *X* (e.g., "space"), named even by a family of different names (Riemann manifold, scheme, Grothendieck *topos*, etc.), but recognized nevertheless under these different names as the foundation of the identity of a hermeneutic trajectory. The Platonic, self-sufficient, and self-identical being of an ideal object is to be replaced by the continuity of a hermeneutic trajectory in which different synchronic and diachronic contexts are discernable and interrelated. In its drift among these varied contexts a mathematical entity (perhaps under different but clearly interrelated names) draws a connected trajectory. And this trajectory is never completed. It always remains within an open historical horizon.

## Patristic Heritage: The *Logoi* of Creation

In conclusion I would like to connect this sketch of a "hermeneutical ontology" with some patristic themes. In the work of the Eastern Church fathers, the idea of a permanent creation, which is more or less a common theological property of different Christian denominations, acquires its peculiar form in the doctrine on the *logoi* of creation. The most important writer for me in connection with the goals of this chapter is St. Maximus the Confessor, a great Byzantium theologian of the seventh century.[7] But even before Maximus, Christian thought about the *Logos* and the *logoi* of things had a rich history of its own. The Christology of the prologue to the Gospel of St. John rapidly developed in the ancient church, especially in the theological school of Alexandria, and not only in explicit relation to concurrent philosophical speculations of Stoic origin. It was also related to the Jewish theology of Philo of Alexandria. A number of ancient Christian writers made use of the Philonian understanding of *Logos* as the true center of the intelligible world. This corresponds to a totality of ideas in the Platonic sense, but there is an essential difference. Philo, as a Jew, sees the *Logos* in terms of a personal deity, and thus the coming together of all ideas in the *Logos* means their coming together in God. Here is the point where Christian writers (such as Pseudo-Dionysius and Evagrius Ponticus) arrive at a different answer. The *logoi* for them are not only more or less static ideas of God lying behind creation (akin to Platonic paradigms, samples for the demiurgic creation of cosmos). According to Evagrius, the *logoi* of *providence and judgment* are to be taken into account.

This means that *logoi* also have to be understood in a *dynamic* sense as "divine and good intentions" for the world. For Maximus, created nature would lose its very existence if it were deprived of its own *energeia*, its proper purpose, and its proper dynamic identity. This proper movement of nature, however, can be fully itself only if it follows its proper goal (*skopos*), which consists in striving for God and a collaboration (*synergia*) with him in fulfilling the logos, or divine purpose, through which and for which it is created. The true purpose of creation is, therefore, communion in divine energy, transfiguration, and transparency to divine action in the world.

During the entire Byzantine Middle Ages, Basil's homilies *On the Hexaemeron* were the most authoritative texts on the origin, structure, and development of the world. Opposing the Hellenistic and Origenistic concept of creation as eternal cyclical repetition of worlds, and affirming creation in time, Basil maintains the reality of a created movement and the dynamism of nature. "Let the earth bring forth" (Gen. 1:24): "this short commandment," says Basil, "immediately became a great reality and a creative logos, putting forth, in a way which transcends our understanding, the innumerable varieties of plans. . . . Thus, the order of nature, having received its beginning from the first commandment, enters the period of following time, until it achieves the overall formation of the universe."[8] While using the Stoic terminology of the *logoi spermatikoi* (seminal reasons), Basil nonetheless remains theologically independent from his nonbiblical sources. For example, he rejects the Stoic idea that the *logoi* of creatures are the true eternal essences of beings, a concept which could lead to the eternal return "of words after their destruction." Basil, as well as Maximus, remains faithful to the biblical concept of absolute divine transcendence and freedom in the act of creation; divine providence, which gave being to the world through the *logoi*, also maintains its existence and fulfills its goal, but not at the expense of the world's own created dynamism, which is a part of the creative plan itself.

In *Questiones ad Thalassium* (Q. II), Maximus clarifies the essential mechanism of this creative dynamics. He writes that although the *logoi* of creatures are fulfilled and perfect in God (not only their being is preserved in the process of permanent creation), but God constantly accomplishes the procession of their potential parts in actual being, rearranging them in a new order. Platonic ideas have no "parts"—they are deprived of any potential constituents that might become actual. That is why they are completely ahistorical. Not so the *logoi* of creation.

Now, coming back to the main subject of my chapter, I would like to stress that the way of being that the hermeneutical ontology ascribes to mathematical objects resembles this patristic philosophy (or theology) of creative *logoi* much more closely than Platonic or even Stoic ontology. In the hermeneutic ontology the main enigma that refers to the true *topos* of "reality," contrary to the alleged autonomous being of concept's extensions before the concept's formation, is the world's "resistance" to conceptual innovations in the face of the common (though culturally diverse) human life-world,[9] in the course of its history. This resistance has its counterpart (which is an unexpected breakthrough in thinking) in a strange fair wind in the sail of a research, an unexpected result of a calculation, or the emergence of new facts that do not fit into the old theories and support the anticipation of a new one. A transcendental concept historically actualizes and rearranges its "potential parts," forming a hermeneutical trajectory.

For the church fathers, this "ultimate reality," which manifests itself as a resistance opposing the arbitrariness of conceptual innovations, is called *God's will and providence in the permanent process of creation*. The hermeneutical analysis of science and, in particular, mathematics allows us to guess how the *logoi spermatikoi* work in Creation. But in itself this is only a step of interpretation, a particular hermeneutical approach, inscribing our research in its proper hermeneutical trajectory.

## Notes

1. Galileo Galilei, "The Assayer," in S. Drake, *Discoveries and Opinions of Galileo* (New York: Doubleday Anchor, 1957), 237f.
2. Cf. P. A. Heelan, "Galileo, Luther and the Hermeneutics of Natural Science," in *The Questions of Hermeneutics*, edited by T. J. Stepleton, 363–75 (Dordecht, The Netherlands: Kluwer Academic Publishers, 1994).

3. P. F. Strawson, *Individuals: An Essay in Descriptive Metaphysics* (London: Methuen, 1959), 10.
4. These ontological positions are presented in full in *Philosophy of Mathematics. Selected Reading*, ed. P. Beņaceraff and H. Putnam, 2nd ed. (Cambridge: Cambridge University Press, 1983).
5. P. Maddy, "Philosophy of Mathematics: Prospects for the 90s." *Synthese* 90, no. 2 (1991): 155–64, here p. 156. Maddy even speaks about "the new consensus" being shaped on this, though internally nuanced, quasiempiricist ground.
6. In connection with this understanding of mathematics that implies a profound ontology of the mathematical entities, I would like to refer to a French school of the philosophy of mathematics, to which, among the others, Jean Petitot and Jean-Michel Salanskis belong. See, in particular, J.-M. Salanskis, *L'Herméneutique formelle. L'Infini—Le Continu—L'Espace* (Paris: editions du CNRS, 1991).
7. For the outline of St. Maximus' theology, see L. Thunberg, *Microcosm and Mediator: The Theological Anthropology of Maximus the Confessor* (Chicago: Open Court, 1995).
8. Basil of Caesarea, *In Hex. hom.* 5, in *Patrologia Graeca*, edited by J.-P. Migne, vol. 29, 1160 D. Quoted in John Meyendorff, *Byzantine Theology* (New York: Fordham University Press, 1979), 133f.
9. A Husserl's term designating the clue concept of the *Crisis of European Sciences and Transcendental Phenomenology*, trans. D. Carr (Evanston, IL: Northwestern University Press, 1970).

## Bibliography

Basil of Caesarea. *In Hex. hom.* 5. In *Patrologia Graeca*. Edited by J.-P. Migne. Vol. 29. Paris: PD Garnier, 1857–66.

Benaceraff, P., and H. Putnam, eds. *Philosophy of Mathematics: Selected Reading*. 2nd ed. Cambridge: Cambridge University Press, 1983.

Galilei, Galileo. "The Assayer." In S. Drake, *Discoveries and Opinions of Galileo*. New York: Doubleday Anchor, 1957.

Hanson, N. R. *Patterns of Discovery*. Cambridge: Cambridge University Press, 1958.

Heelan, P. A. "Galileo, Luther and the Hermeneutics of Natural Science." In *The Questions of Hermeneutics*, edited by T. J. Stepleton, 363–75. Dordecht, The Netherlands: Kluwer Academic Publishers, 1994.

———. "Hermeneutical Philosophy and the History of Science." In *Nature and Scientific Method*, edited by D. Dahlstrom, 23–36. Washington, D.C.: Catholic University Press, 1991.

———. *Space-Perception and the Philosophy of Science*. Berkeley: University of California Press, 1983.

Heidegger, M. *Being and Time*. Translated by J. Macquarrie and E. Robinson. New York: Harper & Row, 1962.

Husserl, E. *Crisis of European Sciences and Transcendental Phenomenology.* Translated by D. Carr. Evanston, IL: Northwestern University Press, 1970.

Kant, I. *Critique of Pure Reason.* Translated by N. Kemp Smith. London: Macmillan, 1958.

Maddy, P. "Philosophy of Mathematics: Prospects for the 90s." *Synthese* 90, no. 2 (1991): 155–64.

Meyendorff, John. *Byzantine Theology.* New York: Fordham University Press, 1979.

Putnam, H. *Representation and Reality.* Cambridge: MIT Press, 1988.

Quine, W. V. *From a Logical Point of View: Nine Logico-Philosophical Essays.* 2nd ed. Cambridge: Harvard University Press, 1961.

Salanskis, J.-M., *L'Herméneutique formelle. L'Infin—Le Continu—L'Espace.* Paris: editions du CNRS, 1991.

Strawson, P. F. *Individuals: An Essay in Descriptive Metaphysics.* London: Methuen, 1959.

Thunberg, L. *Microcosm and Mediator: The Theological Anthropology of Maximus the Confessor.* Chicago: Open Court, 1995.

Torretti, R. *Creative Understanding: Philosophical Reflection on Physics.* Chicago: University of Chicago Press, 1990.

Wiggins, D. *Sameness and Substance.* Cambridge: Cambridge University Press, 1970.

Editor's Introduction to

# Is Mathematics Able to Open the Systems of the Human Intellect?

**Botond Gaál,** Debrecen Reformed Theological University, Hungary

Botond Gaál explores the singular impact of openness in mathematics and theology. Mathematics is a guidepost, the canonical human idea system that allows us to fathom the harmonies of the universe. But it is only in the last two centuries that mathematics has embraced the openness that has revolutionized our capacity to see beyond the boundaries of previous scientific dogma. A similarly true embrace of openness offers Christian theology a route toward invigoration and a new, deeper understanding of human existence.

As with Eastern Europe's societies that struggled within the confines of closed political systems, mathematics long toiled within a set of constrained axioms initially laid down by the Greeks. This system of axioms, based on an overriding sense of the transcendent nature of mathematical objects, cemented the *more geometrico*, a generalized law that all things must fit within the fixed confines of an established pattern. It was not until the nineteenth century and the work of mathematicians like Bolyai and Lobatchewsky that open thinking began to pervade the field. "It was from these seemingly ungraspable and extravisual concepts that a wondrous, breathtakingly new world came into existence." By exceeding the frames of previous conceptual systems, they and subsequent researchers like Cantor and Gödel shattered our concepts of self-reference and infinity, creating an important new context in which to understand God.

Enlightenment theology erred in its acceptance of a mathematical contradiction. Like Kant, the bulk of mainstream theologians imagined that the *theologia naturalis* could be built up from and completely contained within a coherent set of axioms based on foundational theorems. Gaál points out that this is not only mathematically untenable but is "in direct contrast with the basis of Christian belief that considers the Bible, as the source of the revelation of God, to be open."

Botond Gaál calls on theology to learn from the open approach of mathematics and the sciences, an approach that leads to ever-expanding horizons. He warns that if theology chooses closed logics instead, it will simply be left behind by more flexible, expansive, and productive modes of human thought.

P.D.

11

# Is Mathematics Able to Open the Systems of the Human Intellect?

**Botond Gaál,** Debrecen Reformed Theological University, Hungary

## Introductory Ideas

In 1996, Edward Teller delivered a lecture at Debrecen University and categorically declared, “You must understand: modern science means nothing less than that the world is open!” Trained in mathematics and physics, and working as a theologian, I also claim that the structure of the world is of an open nature. Extending this postulate by the premise of inclusiveness, it follows that the human intellect has the same character of openness. Moreover, I regard the congruence of the laws of the universe and of the structure of human thinking as melding into a particular harmony. This is what makes it possible for man to take ever-bolder steps forward in the pursuit of scientific knowledge, the pattern repeated in all fields of science.

A mathematician must recede to the deepest depths of solitude in order to notice relationships that, subsequently, can be equated with universal knowledge or generally accepted truths. This is a lesson clearly taught by the historic development of mathematics. In light of this, it is worthwhile to ponder the following dilemma: how does modern mathematics, through its open structure, serve the acquisition of knowledge pertaining to nature and, simultaneously, the evolution of man's ability to think? Mathematics has proven to be the most effective formal language in formulating descriptions of nature. So it is reasonable to hypothesize that its essential openness may prove to be beneficial not only in scientific discourse but also in other domains, such as the humanities and the practicalities of everyday life. How might such a step forward be taken? In the context of theology, I am led to ask the complementary question: is Christian thinking open enough or not? This issue, I believe, is topical because theologians seem to have let fall into disuse the open approach provided by mathematics, despite having their attention directed toward this possibility by mathematicians. The application of this open approach could have a positive effect on the development of theology as I outline later in this essay.

### The Closed Mathematical World of the Ancients

In about 300 BCE, Euclid collected and gave an overview of all the accumulated mathematical knowledge of his time in the *Elements*. The Greeks are generally taken to have "discovered," "created," and "formalized" mathematics.[1] They *discovered* conceptual logical truths because they believed that those truths already existed in a ready state somewhere within the world of ideas. On the other hand, it is also true that the Greeks *created* mathematics because they were able to acquire new knowledge by using evidence based on the axiomatic system. In the same breath, they *formalized* mathematics because they believed axioms and conclusions derived from them did not necessarily have to be associated with the correlations of the natural world. Eventually they created a field of science whose axiomatic system—according to David Hilbert's twentieth-century terms—was *complete, independent, and free of contradictions*. This is reason enough to praise them, for the truths identified then were as true then as they are now and will continue to be so in the future. Moreover, these mathematical truths can be regarded as scientific truths that are independent of all cultures.

Yet it remains a mystery why the Greeks were unable to harmonize these truths with their knowledge of nature. There may have been something amiss with their approach. Being entirely content with their mathematical method, it would seem that they elevated it to the level of an absolute truth; being unable to imagine anything more perfect, they regarded their method as the most general and unchangeable rule in the cultivation of scientific thought. The term *more geometrico* (the law that all things are to be established on the basis of the geometric model) has its origins here. Thus Euclidean geometry fixed a pattern in almost every field of scientific thought for the next two thousand years. Neither Spinoza, Newton, nor Kant was aware that they were thinking in a closed system.

## The Problem of the Modern Age and the Opening of The Closed World

More than two thousand years later, modern mathematics discovered how to take a step forward. In the 1820s and 1830s, the Hungarian János Bolyai and the Russian Nikolai Lobatchewsky, both mathematicians, concluded that the Greek axiomatic approach led to a closed system of ideas that could and should be changed in the interests of progress. János Bolyai very aptly pointed out that the renowned axiom of parallels had been such an inherent part of Euclid's thinking that it and its influence had precluded thoughts of stepping out of this closed world.[2] However, it was highly desirable that the change ensuing from this stepping out did not result in the loss of established truths. In this respect not even Kant's ideas[3] caused Bolyai to backtrack. From the perspective of the history of science, this might be best described as a "Promethean idea" whereby from the "world of the gods" and its "heavenly fire," Bolyai was able to bring down to earth a small spark that forever changed the world. The idea that he formulated—which even today is revelatory—states that an *infinite* number of lines can be drawn through a point parallel to any given line. This is at least as "Einsteinishly" bewildering as claiming that the velocity of light is constant in any frame of reference. Yet it was from these seemingly ungraspable and extravisual concepts that a wondrous, breathtakingly new world came into existence. In the following years, many capable mathematicians followed Bolyai's and Lobatchewsky's lead and the art of doing mathematics began to flourish anew. The new openness further yielded the establishment of Boolean algebra and this helped attract a slew of mathematicians to the field. Shortly thereafter, the German mathematician Georg Cantor surprised the world with the claim that the human mind is capable of distinguishing between transfinite and absolute infinities.

Until Cantor, it was held that concepts referred to as "absolute" were to be interpreted in terms of the ideal limit of the finite. He pointed out to theologians that although the human intellect was able to grasp the transfinite infinite, it was not able to define God himself as Absolute. Mathematical thinking, moreover, cannot fix God in his ontological nature but can refer to his existence by exceeding its own limits. As Cantor put it:

> To a certain degree the latter is beyond the comprehension of the human intellect inasmuch as it is no longer within the sphere of being mathematically determined. Transfinite infinity, on the other hand, not only utilises a wide range of possibilities in recognising God, but also offers a wealthy and ever-growing space for ideal research. . . . But general recognition is oft times long in coming even if such a revelation could prove to be of extreme value to theologians, it becomes an aid in arguing their case (as for religion).[4]

Cantor inspired more and more mathematicians to examine newer and newer fields. It was somewhat later that the basics of the calculation of probability were introduced, thus opening up new prospects for even more mathematicians. These mathematicians all opened up closed (or supposedly closed) fields and established a new approach for scientific thinking. The same can be said about Kurt Gödel, the twentieth-century Austrian mathematician, according to whose results in logical theory the process of human thinking is open "upward." The work of Alonzo Church and Alan Turing produced a similar result. Mathematicians of the twentieth century not only proved the existence of the open nature of mathematical thinking, thus providing evidence of the open structure of human thinking, but also set their sights on new directions in the spirit of this openness.

## The Discrete and Continuous Mathematics of Our Open World

When the process of resolving dilemmas emerging from axiomatization had ground to a halt, new fields of mathematics offering challenges in research appeared on the horizon. While János Bolyai "*created a new and different world out of nothing*" as far as *Scientia spatii* (i.e., the science of space) was concerned, it was Riemann who continued the opening of space. At the age of twenty-seven, this mathematician had developed a solution for the generalization of Gaussian surface geometry in a higher dimension.[5] This achievement can also be regarded as an opening upward. Physics would ultimately use the new Riemannian mathematics to generate important new understandings of space. But it appears that over the past few decades the number of geometry-related problems in search of resolution has decreased.

Today, mathematical activity is traditionally divided into four categories: creating theories, proving theories, constructing algorithms, and computing.[6] The last of these is more commonly referred to as computer-related science and informatics. Both pure and applied mathematics appear in each of these fields, but it is not always possible to clearly separate the two. In many fields applied mathematics has come to the forefront and has proved useful in supplying better descriptions of nature, natural phenomena, and other sciences (even political science, strangely enough[7]). At the same time, pure mathematics has a host of accumulated tasks waiting upon it in that the natural sciences have evolved in a most rapid fashion also.

Because continuous mathematics cannot describe the events of the "quantum world," it was necessary to develop discrete mathematics that in itself further broadened the imagination of mathematicians. This gave rise to the advent of graph, network, and game theories that represent a certain type of infinity for human cognition. The harmonization of quantum theory with the theory of relativity induced scientists to think in a new mathematical way, in this case resulting in the inception of *string* and *brane* models.

John von Neumann, a renowned Hungarian mathematician, played a preeminent role in the development of twentieth-century mathematics. In describing the mathematical bases of quantum physics,[8] he came to the conclusion that there were no hidden parameters in nature. In principle, there is no limit to cognition, something that mathematicians explain to theologians in the following way: God did not resort to using hidden parameters when he created the world. In discovering all of the above, man could come to admire the openness of the intellect and of the natural world. It was this that gave renewed hope to man in the late twentieth century, and it now serves up new tasks for scientists of the twenty-first century.[9] More and more closed fields have been opened up and worlds unimaginable earlier have been made accessible for scientific research.

### Opening Up the Closed System of Theology

It came to light in the twentieth century that Christian theology could not be built on a system of axioms.[10] Previous to this, many had believed that once a basic theorem was chosen as a foundational point, an entire theological system could be built upon it. This was a result of the influence exacted by *more geometrico* on the theological sciences. Kant expressed similar philosophical views and had consequently gained many followers among theologians. This was the period of the revival and spread of *theologia naturalis*, which, for the most part, ran its course in the nineteenth century.

Theology bore the characteristics of a closed ideological system while contemporary intellectualism communicated open thinking. This was in direct contrast with the basis of Christian belief that considers the Bible, as the source of the revelation of God, to be open. From this it follows that the teachings of the Bible cannot be applied as a system of axioms. Therefore, those who travel in theology should examine such teachings in terms of a scientific and mathematical perspective. It seems evident that if Christian theology truly wants to retain its theological character, it must seek to apply an open way of thinking relevant to its own field in order to comprehend, explain, and interpret dogmas.

This is what the church fathers emphasized when introducing the term *kata physin* (i.e., everything was to be examined according to its own nature). Mathematics shows that the human intellect is infinitely open to the cognition of the created universe. At the same time, faith and religious practice can be enriched by man's effort to understand the revelations of God via the human intellect and by applying these on a daily basis.

Already there have been some benevolent warnings emanating from mathematics. It is my conviction that theology must learn from the open approach employed by the exact sciences. In neglecting to do this, theology will not be able to yield any tangible results. Moreover, ecumenical efforts may also produce nothing other than a hollow ring. Therefore it is to be recommended that all denominations apply open theological thinking in identifying those clauses that cap or close their system of beliefs and also those clauses that uphold an ideology or inflexible dogma in the name of "scientificity." As long as theology remains mired down, all the other sciences will leave it in their wake. This requisite can no doubt be regarded as the *conditio sine qua non* of all ecumenical efforts in the twenty-first century.

## Where Can Closed Systems Be Opened in the Present-day World?

The enquiries of mathematics and theology as two theoretical subjects can be practically applied. Although experience shows that a system does not have to be axiomatically constructed in the mathematical sense, it can become closed. Observing the systems working around us, we can find several closed systems. Certainly since these are active, "alive," moving systems, we can examine numerous cases in our proximate environment. A political social system can be closed if its leaders rule it in a totalitarian way, if they apply ideological coercion and do not permit the creative development of the people living there. But a political system also becomes closed if absolute licentiousness is allowed in it, when public life falls into anarchy and chaos impedes the formation of order.

Likewise, the behavior of people inside a social system, their conduct in their moral world, can be closed or open. It is closed if a person regards individual autonomy an absolute rule and subordinates everything to it in his life. At the same time there is the possibility of openness, but it has its own order, when the person acknowledges heteronomy as well. In the same way, if we examine the question of church organization, we also find such phenomena that make it either closed or open. And in the field of church doctrine systems, we can find very extreme examples of closedness. But where churches ensure openness, there the Christian view of life manifests itself more powerfully.

Finally, it is worthwhile to examine the universities. In every society the universities determine the whole educational system and its intellectual life. A university works well that ensures the freedom of conscience in the cultivation of the sciences. Man is a being "open upwards," made for creation, one of whose greatest responsibilities in every society and organization is to ensure openness for creation. That is why the lesson of the development of sciences can be useful in the life of a society and that is why the sciences result in a higher spirituality in the human being.

## Notes

1. Compare John D. Barrow, *A fizika világképe* [*The Worldview of Physics*]. (Budapest: Akadémiai Publishing, 1994), 64.
2. Compare Zoltán Gábos, "Mit adott a fizikának Bolyai János?" ["How Did János Bolyai Enrich Physics?"], in *Bolyai Emlékkönyv* [*Bolyai Commemorative Volume*]. Budapest: Vince Publishing, 2002), 269. "An axiom has a specific and separate role in Euclid's system, since the statement it consists of emphasizes and fixes its Euclidean nature. At the same time, it represented a stable element which precluded stepping out of the Euclidean system. Removing the 'barrier' opened up a path to a new, logically viable geometry and, at the same time, a new model of space."
3. Bolyai thought the following about Kant's ideas of space: "The otherwise honourable and clever Kant insisted on his groundless and twisted theorem that space . . . was not self-consistent but only an idea or a frame for our visions[!]" as it was quoted by Zoltán Gábos, "Mit adott a fizikának Bolyai?" in *Bolyai Emlékkönyv*, 274.

4. ELTE, *Filozófiai Figyelő*, Budapest (1988/4), 82–83.
5. It is quite interesting to discover how the geometry developed by Riemann came into being. Riemann submitted an application for habilitation examination at Göttingen University in 1853. Traditionally, proposals for three lectures had to be submitted. Riemann had prepared only the first two because the habilitation committee generally always asked to hear the first one. But for once it happened differently. Gauss, who was also a member of the committee, wanted to hear the third lecture. This is why Riemann wrote to his younger brother that he was in difficulty. Eventually he managed to prepare the lecture and this habilitation lecture gave rise to a world-famous discovery, something which gave a lot of work for geometers following Riemann in time. Compare János Szenthe, "Relationship between Hyperbolic Geometry and Riemann's Geometry," in *Bolyai Emlékkönyv*, 308–9, 312.
6. Compare András Prékopa, "Gondolatok a Matematikáról" ["Ideas of Mathematics"]. *Confessio* (1998/1), 9.
7. It was to a large part the mathematical development of the game theory that enabled the Americans to foretell—with quite high accuracy—how the Soviet politicians would react to certain issues. As yet only very few of the details are known, but it became possible that one of the parties at the negotiating table could predict the answer to his question that would be forthcoming. It also made it easier to prepare for such negotiations. These interesting events took place in the second half of the twentieth century.
8. His famous work is entitled "Mathematische Grundlagen der Quantenmechanik."
9. A book has been published under the title *Opening Up a Closed World* by the author of this essay (Debrecen, Hungary: István Hatvani Theological Research Center, 2007). In this book he has taken ten examples from humanity's cultural history, which demonstrated well how closed systems of thought have to exist, and the manner their unlocking has taken place. The chapters are as follows: "The Religious Situation of the Ancient World—Significance of the Jews and Early Greeks"; "The Greek *More Geometrico* Period"; "Jewish Monotheism and the Christian Trinitarian Perspective"; "Ptolemy Closes, Copernicus Opens"; "Europe Establishes a New Mathematics"; "János Bolyai: Out of Nothing I Have Created a New and Different World"; "Axiomatization and the Upward Opening Infinite World"; "A Mathematician Offers Religions a Change of Perception"; "Theology and a New *More geometrico* Perception"; "The Continuous and Discrete World."
10. The theology of Karl Barth provides the best evidence of this.

## Bibliography

Barrow, John D. *A fizika világképe* [*The Worldview of Physics*]. Budapest: Akadémiai Publishing, 1994.

*Bolyai Emlékkönyv* [*Bolyai Commemorative Volume*]. Budapest: Vince Publishing, 2002.
ELTE. *Filozófiai Figyelő*, Budapest (1988/4), 82–83.
Gaál, Botond. *Opening Up a Closed World*. Debrecen, Hungary: István Hatvani Theological Research Center, 2007.
Prékopa, András. "Gondolatok a Matematikáról" ["Ideas of Mathematics"]. *Confessio* (1998/1), 8–18.

## Editor's Introduction to
# On the Role of Transcendence in Science and Religion

**Ladislav Kvasz,** Catholic University in Ruzomberok, Slovakia

Ladislav Kvasz boldly proclaims that a mature relationship between science and religion can be found on the level of transcendence. Delving deep into the analytic construct pioneered by Ian Barbour, Kvasz emerges with a novel concordance between scientific and theological schema. After a careful appraisal of Kuhn's *Structure of Scientific Revolutions*, Kvasz is able to reorganize typologies of science and religion in such a way as to allow Barbour's fourfold classifications to neatly dovetail with independent frameworks for each. This ambitious work sets the stage for a larger reconceptualization of science and religion with the potential for far-reaching impact.

Ian Barbour's work has become the touchstone of our field over the past decade and a half. His appeal lies largely in a simple categorical scheme that divides analytic modes in science and religion into those featuring conflict, independence, dialogue, and integration. Kvasz embraces this typology and offers brief commentary on each mode and its appropriation by particular intellectual constituencies. Most importantly, he sees the four modes as "disclosing four levels of complexity on which the system of science interacts with the system of religion."

These four levels of complexity neatly match four distinct meanings implicit in Kuhn's ambiguous term *scientific revolution*. Kvasz introduces four specific kinds of change that each characterize revolutionary overthrow in scientific systems but whose distinctiveness has too often been overlooked in analyses of Kuhn's notion of "paradigm." An irrevocable change in the language of science, its reformulation, certainly constitutes an overthrow, at least a minor revolution. But the change in a theory's ontological foundations, an objectivization, is more momentous (changes of this type have often been later overthrown themselves). But sweeping, deep revolutions result from the tensions between the new objects introduced by objectivization and the conceptual framework of preexistent science. Re-presentation demolishes the old and rebuilds a new conceptual structure. To these three modes, Kvasz adds a fourth layer of complexity, the idealization that not only brings into being an original structure but introduces a worldview so new as to create a space for entirely novel scientific epistemologies.

It appears that idea systems are in conflict when the analyst is trapped in a level of language, where propositions collide without their underlying ontologies coming to the fore (Kvasz' level of reformulation). Objectivization allows for ontological shifts on their own terms—the conceptual systems that emerge are independent. And the re-presentation of science takes place in boundaries, places where the ambit of science shifts and overlaps others, making it the level of dialogue. In bridging each group, Kvasz offers a roadmap linking a Kuhnian analysis to the familiar one of Barbour, but he leaves the toughest link for last.

A classic typology of religious doctrine is found in Lindbeck's 1984 book *The Nature of Doctrine*. Kvasz reorganizes the ideas presented there to build a fourfold typology that, he holds, is ultimately a "classification of the different levels or layers of the language of theology." Three of these are explicit in Lindbeck's work: the propositional, linguistic, and expressive approaches. As poles in a theological spectrum, propositional thinking (like biblical literalism) conduces to conflict, Wittgensteinian and other linguistic approaches suggest independence, and expressive currents like those present in nineteenth-century theology are open to dialogue. The fourth, coined by Kvasz himself, is a "function of transcendence."

It is transcendence that makes a frame within which to understand Kvasz' overarching thesis. Religion's function is to provide a continuous thread back to the moment at which our consciousness emerged. It was there that we developed the capacity to constantly bring a world into being by perceiving it, and so it was the moment "when we came into being as men." It is in the parallels between this constant transcendence and idealizations—the open processes of new beginnings in science—where we will finally build a mature relationship between science and religion.

P.D.

12

# On the Role of Transcendence in Science and in Religion

**Ladislav Kvasz,** Catholic University in Ruzomberok, Slovakia

## Introduction

The different relations between science and religion present a complex problem having historical, political, sociological, cultural as well as cognitive dimensions. All these aspects of the problem have been discussed in the literature and the conflicting views on the relation between science and religion are often the result of emphasizing some aspects while neglecting others. The aim of the present chapter is to outline an approach to the study of the relations between science and religion that would allow us to respect the complexity of these relations and, at the same time, to offer a better orientation among them. As a starting point we will take the influential classification of the ways of relating science and religion crafted by Ian Barbour. Assuming that between science and religion there really was conflict, independence, dialogue, and integration, we can pose the following question: *what levels of complexity must there be in science as well as in theology in order that the four types of relation between them become possible*? It turns out that in the philosophy of science as well as in theology there are interpretations that represent science (Kvasz 1999) and theology (Lindbeck 1984) on such a level of complexity that is necessary with sufficient complexity to apply Barbour's classification.

In his classic *Religion in an Age of Science* (Barbour 1990), Barbour proposed a classification of the ways of relating science and religion. According to this classification, we can divide the views on the relations between science and religion into four types:

The first view understands the relation of science and religion as a *conflict*. Scientists who promote a *materialistic worldview* and see religion as an outdated system of superstitions hold this conviction most often. On the religious side the supporters of biblical literalism embrace the conviction that there is a conflict between science and religion. Both materialistic scientists and advocates of a literal interpretation of biblical texts consider the propositions of science and of religion to be competing assertions about one and the same reality.

The second view understands science and religion as *independent* pursuits. According to the proponents of the independence thesis, science is based on experimental method, its language is mathematical, and its goal is the growth of knowledge. Religion, in contrast, uses a hermeneutical method, its language is metaphorical, and its goal is to change human existence. In this way the factual gets separated from the existential. *Existentialist* theologians as well as scholars using *linguistic analysis* embrace this view.

The third view understands the relation of science and religion as (having a potential for) a *dialogue*. According to the proponents of this view, there are points of contact and correlation between science and religion that can be developed into a mutual dialogue. This conviction is attractive to historians, philosophers, and theologians who study the *presuppositions of the scientific discourse* such as the intelligibility of the world, *limit questions* such as the ethical issues connected with the use of the results of scientific research, and the *methodological parallels* between science and religion. This is also the view held by religious scientists who discern a *sacred order in nature*.

The last, fourth view understands the relation of science and religion as *integration*. Barbour distinguishes three areas for integration. The first is *natural theology* according to which the existence of God can be inferred from the evidence of design in nature. The second is the *theology of nature*. Barbour characterizes it as the conviction that even though the main sources of theology lie outside science, it must, nonetheless, have intellectual credibility in modern society and its doctrines must be consistent with scientific evidence. Probably the most outstanding representative of theology of nature was Teilhard de Chardin. The third area of integration is represented by *process philosophy*. It denies the duality of matter and mind. It understands nature as a permanent process of change, with God as the ultimate source of novelty and order.

Barbour's classification manifests an ordering by increasing affinity between science and religion. If we look at Barbour's four categories from the viewpoint of philosophy of science, we find that in each of the four kinds of interaction a *different aspect of science* is involved. Science enters into the first category, the category of *conflict*, in the form of a closed system of "proven truths," which can be enwrapped in scientific materialism, positivism, or some other kind of scientism. In the category of *independence*, the stress is put on questions of method. The role of scientific method is to maintain the openness of science, thus it shifts the focus from "proven truths" to new discoveries. The category of *dialogue* concerns limit questions, i.e., questions that defy strict methodological treatment and even unambiguous formulation. These are the questions where science touches the boundaries of its discourse. And finally, in the case of *integration*, science loses its integrity and its dependence on the constitution of the human mind, and being in general comes to the fore.[1]

Barbour's classification can be thus interpreted as disclosing *four levels of complexity* on which the system of science interacts with the system of religion. When we accept this interpretation, we will be able to distinguish in each of Barbour's four categories its scientific and theological pole. Then we can interpret a type of interaction between science and religion as an interaction between the corresponding poles. Surprisingly, the four aspects of the scientific discourse that enter into the particular categories of Barbour's classification correspond with the levels of change described in the classification of scientific revolutions (Kvasz 1999). It was the realization of this particular correspondence that led me to look for a similar theory linking to theology's poles. But before turning to theology, let me make the correspondence between the theory of scientific revolutions and Barbour's theory more explicit.

### A Refinement of Kuhn's Theory of Scientific Revolutions

Thomas S. Kuhn introduced the notions of scientific revolution, paradigm, and scientific community in his *The Structure of Scientific Revolutions* (Kuhn 1962). Kuhn's theory has often been criticized because his notion of paradigm is vague and has many meanings. As an attempt to answer this criticism, I proposed a classification of scientific revolutions (see Kvasz 1999). I tried to show that there are at least three different kinds of revolutions, which can be called *idealization*, *re-presentation,* and *objectivization*. These three kinds of scientific revolutions can be expanded by a fourth kind of change called *reformulation* (this is characteristic for normal science). The perceived vagueness of Kuhn's notion of paradigm is then the consequence of mixing of these different kinds of scientific revolutions.

Even though the classification of scientific revolutions originated in a different context than Barbour's classification, its four kinds of scientific change show a close correspondence with the scientific poles of Barbour's four categories. Since the notion of scientific revolution has many different meanings and is used mostly in a sociological context, I suggested introducing the notion of an *epistemic rupture*, which could be used in a more neutral way (Kvasz 1999, 208). The term *epistemic rupture* represents the epistemological aspect of the changes that occur in the course of a scientific revolution (i.e., phenomena like the possibility of translation between the different paradigms, their methodological, logical, or linguistic incompatibility) independently of the way in which the scientific community reacted to these epistemological changes (by rejecting the old paradigm; acknowledging the parallel existence of more paradigms; or by incorporating a fragment of the old paradigm into the new one).

A *reformulation* is an irreversible change of the language of a particular theory, which does not alter the theory's conceptual framework.

Examples include the discovery of the planet Uranus, the experimental determination of the speed of light, and Planck's derivation of the law of black body radiation. After the discovery of Uranus, a new name was introduced into the language of astronomy, after the determination of the speed of light a new constant was introduced into physics, and after Planck's derivation, thermodynamics was enriched by a new law.

An *objectivization* is a change of the conceptual framework of a particular physical theory that is the consequence of a change of the theory's ontological foundations. Most often it involves the postulation of a new kind of substance that has hitherto unknown or unusual properties. Examples of objectivization include the postulation of phlogiston by Georg Stahl, the postulation of ether by James Maxwell as the medium transmitting electromagnetic waves, the postulation of atoms by Ludwig Boltzmann as an explanation of thermodynamic phenomena, or the introduction of the light quantum by Albert Einstein in his theory of the photoelectric effect. In all of these cases, the existence of a new kind of object or of a new substance was postulated and it was added to the building blocks of the universe. From the epistemological point of view, it is not important whether the newly postulated substance remained in physics as a permanent component (as in the case of atoms and quanta) or was discarded (as in the case of phlogiston and ether). The important thing is that the ontological structure of a physical theory was changed.

A *re-presentation* is a change that is usually the result of the tensions between the newly introduced objects, postulated in an objectivization, and the old conceptual framework of the theory. The result is a radical change of the whole conceptual structure. Perhaps the best-known re-presentation happened during the Copernican revolution that totally demolished the Ptolemaic-Aristotelian system with its theory of natural place. Another well-known re-presentation accompanied the Einsteinian revolution with its replacement of Newtonian space and time by the general theory of relativity. Similarly, the quantum revolution radically changed our understanding of matter.

An *idealization* is an even more radical change of the structure of scientific knowledge than re-presentation. Such an idealization occurred in the seventeenth century and culminated in the work of Isaac Newton—the change that separates the ancient ideal of theoretical knowledge, represented by Euclid's *Elements*, from the modern ideal of experimental science, represented by Newton's *Principia*. Despite their many differences, classical mechanics, field theory, and quantum mechanics are all constructed according to a common scheme. According to all three of them, the core of the physical description of reality consists in the determination of the state of a system, of the *dynamic equations* (determining the temporal evolution of the state), and of the *symmetries of the system*. The first theory that introduced the idea of the description of a physical system using the notions of state, dynamic equations, and symmetries was Newton's mechanics.

## Correlation between Kuhn's Refined Theory and Barbour's Typology

It is interesting to realize that Barbour's four ways of relating science and religion are in close correlation with the four kinds of epistemic ruptures described in the previous section. The ruptures occur at the very same levels of complexity of the system of science that define Barbour's classification.

It is easiest to demonstrate the connection between the level in the system of science at which the *conflicts* between science and religion occur and the level at which *reformulations* happen. If we try to reduce science to a system of propositions (as positivism often does), the only changes we are able to detect are reformulations, because the more complex issues of ontology are out of sight. And since religion is then seen only as an alternative set of propositions, *conflict* is the inevitable relationship.

Similarly straightforward is the relation between *independence* and *objectivizations*. Objectivizations are closely related to the conceptual framework of scientific theories, i.e., the framework where the semantics of their fundamental categories, their system of legitimate methods, and their ontological foundations are fixed. A conceptual framework is highly interconnected but also relatively self-sustained. If we see science as a system of propositions embedded in a particular conceptual framework, i.e., if besides its propositions, we also take into account its methods and ontological assumptions, we find ourselves on the level where independence from religion is most obvious.

It is not difficult to show the connection between the level in the system of science at which the possibility of a *dialogue* between science and religion opens and the level at which *re-presentations* happen. In recent work on re-presentations in physics, I characterized the particular re-presentations as shifts of the boundaries of the language of science. According to this characterization, each representation[2] has peculiar logical and expressive boundaries and the re-presentations happen on these boundaries. By means of a radical change in language, they surpass the logical and expressive boundaries of the theory. Barbour finds the potential for the dialogue on the boundaries of the scientific discourse on the very same level in the system of science on which re-presentations occur.

I will turn to the question of the connection between *integration* and *idealizations* in the last sections of this chapter. It is the most interesting form of relation between science and religion but also the most problematic. While Barbour was able to illustrate the first three categories of his classification with episodes from the standard history of science, in the case of integration he cites natural theology, theology of nature, and process philosophy—theories whose relation to normal science (in Kuhn's sense) is not very clear. More standard examples from the history of science illustrating integration are available in our correlation between Barbour's theory and the classification of scientific revolutions. The fact that integration is linked with idealization enables us to select a few episodes as candidates for integration from the waste area of the history of science. But before we make this move, let us introduce the theological pole of Barbour's classification.

## A Reordering of Lindbeck's Theory of the Nature of Doctrine

George Lindbeck in *The Nature of Doctrine* (1984) described three approaches to religious doctrines. These approaches can be brought into correlation with the theological poles of the first three categories of Barbour's classification. The only thing we have to do is to change the order of Lindbeck's approaches. In this way we obtain the following list:

*The propositional approach* can be characterized, according to Lindbeck, by the conviction that religious doctrines are true propositions about objective reality. This approach to religious doctrines represents the theological pole of the first kind of relations between science and religion in Barbour's classification, namely *conflict*. Barbour illustrates it with the example of biblical literalism. If we understand religious doctrines or passages in the Bible literally and interpret them as true assertions about objective reality—the same reality that is studied also by science—we end in a conflict.

*The linguistic approach* interprets the doctrines not as true propositions about the world but rather as rules of the religious discourse. This approach was inspired by the philosophy of Ludwig Wittgenstein, and it seems to be in correspondence with the theological pole of the second kind of relations between science and religion in Barbour's classification, namely *independence*. Barbour explicitly stated that, in the case of their independence, science and religion are usually understood as different languages. If we interpret the religious doctrines as rules that constitute the religious discourse, this discourse becomes independent from the discourse of science.

*The expressive approach* is, according to Lindbeck, characteristic of the theological currents of the nineteenth century. In these currents, doctrines were viewed as symbolic expressions of religious experience, that is, of an internal experience with the divine. Bringing this approach into a correlation with the theological pole of the third kind of relations between science and religion in Barbour's classification, namely *dialogue,* cannot be achieved as directly as in the previous two cases. Barbour cites the limit questions in science and discerning the sacred order in nature. If we agree that one of the roles of religious symbols is to remind us of the limits of the scientific discourse and that religious experience can open us toward the perception of the sacred order in nature, then the connection between the expressive approach in theology and Barbour's category of dialogue is acceptable.

## On Correlation between Lindbeck's Theory and Barbour's Typology

Despite its preliminary character, the correlation between Barbour's classification and Lindbeck's theory is surprising. After introducing into Lindbeck's typology the order stemming from Barbour's classification, the internal logic of Lindbeck's system becomes apparent. While the *propositional approach* anchors religious doctrines in "external" reality (understanding them as propositions about the external world) and the *expressive approach* anchors them in "internal" reality, understanding the doctrines as symbolic expressions of internal religious experience, language is the medium that connects these two realities. The meanings of the expressions of a language belong to the "internal" world, their referents to the "external." Thus, language itself is on the boundary of these two realms. Therefore our modified arrangement of Lindbeck's approaches seems to be more natural than the original one that we find in Lindbeck's own work.

Nevertheless, one problem remains unaddressed. While Barbour's classification contains four kinds of interaction between science and religion, Lindbeck recognizes only three approaches to the nature of doctrine. Let us now focus on this problem.

If we wish to correlate the fourth kind of scientific revolutions, idealizations, with its theological counterpart, we have to find a fourth approach to the nature of religious doctrines that would fit into Lindbeck's theory. It is surprising that what we are looking for is, at least in an implicit form, already present there. By changing the order of the approaches of Lindbeck's classification, the linguistic approach was bracketed between the propositional and the expressive ones. Due to this rearrangement, the linguistic approach received clearer boundaries and it became obvious that Lindbeck's book contains some hints of a fourth approach. These hints simply did not fit into the new place (between the "external" and the "internal") to which the linguistic approach has been shifted.

Lindbeck generally compares his postliberal theology to a natural language, and he suggests interpreting religious doctrines as rules of grammar of the religious discourse. This can be illustrated by several passages from Lindbeck himself: "Religions are seen as comprehensive interpretive schemes, usually embodied in myths or narratives and heavily ritualized, which structure human experience and understanding of self and world" (Lindbeck 1984, 32). "A religious system is more like a natural language than a formally organized set of explicit statements, and that the right use of this language, unlike a mathematical one, cannot be detached from a particular way of behaving" (Lindbeck 1984, 64). In these and similar passages, religious doctrines are interpreted as rules, i.e., as something that forms our life rather than saying something about the world.

However, in addition to such passages we also find in Lindbeck something altogether different. First he quotes Wilfrid Sellars according to whom "the acquisition of a language is a jump which was *the coming into being of man*," and then writes: "The Christian theological application of this view is that just as an individual becomes human by learning a language, so he or she begins to become a new creature through hearing and interiorizing the language that speaks of Christ" (Lindbeck 1984, 62). If we situate this last quotation among the others introduced earlier (as it is in Lindbeck's book), it appears that they all speak about the same thing—about language as a system of interpretive schemes, rules of grammar, or second-order assertions. But the coming into being of man cannot be described using interpretive schemes or rules of grammar because there is nobody who could use these schemes or rules. When a child *jumps to consciousness*, this undoubtedly happens with the help of language. The parents speak to the child in more or less articulated ways. But the basic premise, that the use of interpretive schemes or rules of grammar have the power to initiate this jump, is mistaken. Language as a system of interpretive schemes and rules of grammar presupposes consciousness; therefore it cannot explain the *birth* of consciousness.

I am not denying that language plays a fundamental role in the birth of consciousness. But this is not language understood as a system of interpretive schemes and rules of grammar. Of course, all three approaches to doctrine described by Lindbeck are in one way or the other related to language. One of language's most important functions is to enable us to pronounce propositions about the world, as is its capacity to allow us to use metaphors and parables for symbolic expression of our internal experience. Therefore not only the linguistic approach, but the whole classification proposed by Lindbeck is based on the analysis of different functions of language. We can interpret his scheme as a *classification of the different levels or layers of the language of theology* (as our classification of scientific revolutions can be seen in this way as describing the different levels or layers of the scientific language). The language of theology, just like the language of science, has several levels. We can discriminate a *propositional* level, a *normative* level (the level to which the linguistic approach to doctrines is based), and a *symbolic* level (on which the expressive approach is based). It is language that enables us to assert something about the world, to discuss norms of our behavior, and to express our experience in a symbolic way.

Nevertheless, it seems that besides the three levels that form the basis of Lindbeck's classification, language has a fourth level—the ability to detach man from immersion in his environment, to create a distance from it and to constitute man as a conscious being. This function of language I suggest calling the *function of transcendence*. When parents speak to their small infant, they are using this magical power of language that enables them to engross the attention of the child and liberate it from captivity in its momentary sensory perceptions. But this ability to engross and liberate, this ability to *transcend* the given, is founded neither on the propositional layer of language (the parents can address the infant with true as well as false propositions, it does not understand them anyhow), nor on the regulatory or symbolic layer (it does not matter whether the parents follow the rules of grammar or violate them). Most often the parents use meaningless sounds and it works nonetheless.

### The Role of Transcendence in Religion and in Science

We have reached the point where I can express a hypothesis about religion, a hypothesis that is the central thesis of this whole chapter: the main purpose of religion is neither in its propositional function (to offer us a true description of the transcendent being of God), nor in its linguistic function (to provide us with a linguistic framework for speaking about the things most important for man), nor in its expressive function (to offer symbols for the expression of the internal experience with the divine). The main function of religion is rather to keep continuity with the moment when we *came into being as men*. Religious rites, rituals, and ceremonies are symbolic recapitulations of this event, of the event of our coming into being, our origin, and our creation. Thus, religious rites, rituals, and ceremonies can be seen as means for resuming contact with the constitutive center of our being. The beginning described by religions is the beginning of our consciousness and thus also the birth or creation of the world as something to which we have conscious access.

An interpretation of religion as of that which created us as conscious beings, and an interpretation of the history of religion as the history of expansion of human consciousness, may not be acceptable from a theological point of view. That remains to be decided after development and exploration of a more comprehensive theory. At present I would like to draw attention to only one consequence of this hypothesis for the relation of science and religion. If we accept the view that the main function of religion is the function of transcendence, it becomes obvious that if religion wants to engage in a mature relationship with science, we must understand science in the state of its becoming. A full-fledged relationship between science and religion is possible only in those moments when the system of science is in living contact with its own layer of transcendence. And this is precisely what happens in the process of idealization.

## Notes

1. I. Barbour, *Nature, Human Nature, and God* (Minneapolis: Fortress, 1997), 35.
2. The term *re-presentation* (written with a hyphen) stands for an epistemic rupture, i.e., a fundamental epistemic change. Such a rupture separates two *representations*, i.e., two ways how we see and interpret reality.

## Bibliography

Barbour, I. *Nature, Human Nature and God*. Minneapolis: Fortress Press, 2002.

———. *Religion in an Age of Science*. New York: Harper Collins, 1990.

———. *Religion and Science*. San Francisco: Harper, 1997.

Kuhn, T. S. *The Structure of Scientific Revolutions*. Chicago: Chicago University Press, 1962.

Kvasz, L. "The Invisible Dialog between Mathematics and Theology." *Perspectives on Science and Christian Faith* 56 (2004): 111–16.

———. "The Mathematization of Nature and Newtonian Physics." *Philosophia Naturalis* 42 (2005): 183–211.

———. "On Classification of Scientific Revolutions." *Journal for General Philosophy of Science* 30 (1999): 201–32.

———. *Patterns of Change: Linguistic Innovations in the Development of Classical Mathematics*. Basel: Birkhauser, 2008.

Lindbeck, G. *The Nature of Doctrine*. Philadelphia: Westminster Press, 1984.

Russell, R. J., W. R. Stoeger, and G. V. Coyne, eds. *Physics, Philosophy, and Theology: A Common Quest for Understanding*. Vatican City: Vatican Observatory, 1988.

# Contributors

**Dr. Grzegorz Bugajak** is associate professor at the Institute of Philosophy (Section of the Philosophy of Nature) of the Cardinal Stefan Wyszynski University in Warsaw. He is a member of the editorial team of *Studia Philosophiae Christianae* (a scholarly periodical for philosophy), the European Society for the Study in Science and Theology, and the Polish Philosophical Society.

**Dr. Alexei Chernyakov,** PhD in mathematics (St. Petersburg State University), Doctor of Philosophy (Free University of Amsterdam), Doctor of Philosophy (Dr. Hab., Russian State Humanities University, Moscow), is the chairperson of the Metanexus/LSI program "St. Petersburg Education Center for Religion and Science" (SPECRS) and the head of the Department of Philosophy at St. Petersburg School of Religion and Philosophy. He is particularly interested in ancient philosophy, phenomenology (Husserl, Heidegger, Levinas, etc.), and Greek patristic science and religion. Prof. Chernyakov is also the author of two books and a number of articles, published in leading international journals, dedicated to the problems of mathematics, philosophy, and the Russian Orthodox tradition.

**Dr. Pranab Das** is chair and professor of physics at Elon University. Having focused his scientific research on chaos theory and nonlinear dynamics, he is presently involved in the rich interdisciplinary dialogues that arise from the intersections of science and society. He is executive editor of the International Society for Science and Religion's Library Project and the leader of the Global Perspectives on Science and Spirituality program. During the past several years he has worked closely with scholars from around the world to bring their insights to a Western audience and to foster excellent research in their unique approaches to some of the key questions of our times.

**Dr. Botond Gaál** is professor of systematic theology at Debrecen Reformed Theological University. He earned a diploma in mathematics, physics, and theology. He studied theology at New College, Edinburgh, Scotland, and was invited to research in the relationship between science and theology at the Center of Theological Inquiry at Princeton, New Jersey, United States. He holds a doctorate in divinity, made scientific investigation in the roots of the Reformation, and founded a special institute for the study of science and theology in Debrecen. Gaál is known as a specialist in James Clerk Maxwell's and Michael Polanyi's significance in European civilization and he holds a special title, doctor of the Hungarian Academy of Sciences.

**Dr. Jiang Sheng** is the founder, director, and presently professor at the Institute of Religion, Science and Social Studies, at Shandong University, and specially engaged professor at Fudan University, Shanghai. He is a member director of the Chinese Society of Religious Studies, a member director of the Chinese Confucius Foundation, member of the Japanese Society of Taoistic Research, anf founding executive of the Hong Kong Taoist Culture and Information Center and Mnbre du Conseil Scientifique de l' Universite Interdisciplinaire de Paris. As the leading scholar in the study of Daoism and science, he has been received grants twice—in 1998 and 2006 by China's National Social Sciences Foundation—for the national major project of "History of Science and Technology in Taoism," and has founded an international group for its purposes. Jiang has been a visiting scholar at Harvard University and a research professor at the University of Virginia. He has received numerous honors and academic prizes and is a winner of both Global Perspectives on Science and Spirituality (GPSS) awards. He has been engaged as Mt. Tai Distinguished Chair Professor at Shandong University by Shandong Province since 2005 and has served as Chair Expert of Shandong Province Center for Asian Studies since 2003.

**Dr. Ilya Kasavin** is a correspondent member of the Russian Academy of Sciences, a professor at the Institute of Philosophy, Russian Academy of Sciences, and professor at the Russian State University for Humanities. He is also founder and general secretary of the Centre for the Study of German Philosophy in Moscow as well as founder and editor-in-chief of *Epistemology & Philosophy of Science*, the journal of the Institute of Philosophy of the Russian Academy of Sciences.

**Dr. Heup Young Kim** is professor of systematic theology at Kangnam University in South Korea. He is a former dean of the College of Humanities and Liberal Arts, the Graduate School of Theology, and the University Chapel. He earned a BSE from Seoul National University, an MDiv and ThM from Princeton Theological Seminary, and a PhD from the Graduate Theological Union. He has carried out extensive research in the area of science and religion, been a visiting scholar at the Center for Theology and the Natural Sciences (Graduate Theological Union), a senior fellow at the Center for the Study of World Religions (Harvard University), and is one of the founding members of the International Society for Science and Religion. Professor Kim has received numerous honors and awards, including the John Templeton Foundation Research Grant (2004–5), Distinguished Research Professor Award, Kangnam University (2003), and Most Distinguished Research Professor Award, Kangnam University (2003).

**Dr. Ladislav Kvasz** has been employed at the Faculty of Mathematics and Physics of Comenius University since 1986. In 1993 he won the Herder Scholarship and spent the academic year 1993–94 at the University of Vienna studying the philosophy of the Vienna Circle and Ludwig Wittgenstein. In 1995 he won the Masaryk Scholarship of the University of London and spent the 1995–96 academic year at King's College London studying the philosophy of Imre Lakatos. In 1997 he won the Fulbright Scholarship and spent the summer term of the 1998–99 academic year at the University of California at Berkeley working on Husserl's theory of the Galilean revolution. In 2000 he won the Humboldt Scholarship and spent 2001 and 2002 at the Technical University in Berlin studying the epistemological background of the scientific revolution. He is currently a member of the Union of Slovak Mathematicians and Physicists (JSMF) and the Slovak Philosophical Society (SFZ). Since 2004 he has been the director of the Center for Interdisciplinary Studies at the philosophical faculty of the Catholic University in Ruzomberok.

**Dr. Anton Markoš** holds a PhD in biology and physiology from the Faculty of Sciences at Charles University in Prague. He has been teaching cell physiology and developmental biology since 1972. Between 1994 and 2002 he was a member of the Center of Theoretical Study, a transdisciplinary body formed by the Charles University and the Czech Academy of Sciences. Since 2002 he has been the head of the Department of Philosophy and History of Sciences at the Faculty of Sciences.

**Dr. Sangeetha Menon** is a faculty professor at the National Institute of Advanced Studies, Bangalore, India. She joined NIAS in 1996. A gold medalist and first-rank holder for postgraduate studies, she received a University Grants Commission fellowship for her doctoral studies for five years. Menon has been working in the area of consciousness studies for over fifteen years. Her core research interests include Indian ways of thinking and experiencing and current discussions on "consciousness." Her work and experience involves a combination of scientific engagement, creative interests, and spiritual pursuit. Menon has authored one book: *The Beyond Experience: Consciousness in the Gita* (Srishti, 2008); and coedited four books: *Consciousness, Experience and Ways of Knowing* (NIAS, 2006); *Science and Beyond: Cosmology, Consciousness and Technology in Indic Traditions* (NIAS, 2004); *Consciousness and Genetics* (NIAS, 2002); and *Scientific and Philosophical Studies on Consciousness* (NIAS, 1999). The book she has coauthored with Swami Bodhananda, *Dialogues: Philosopher Meets Seer* (Srishti Publishers, 2003), is a set of nine dialogues with her Guru on sociocultural issues of contemporary importance. Currently she is writing two books on self transformation and spiritual agency. Her website is www.samvada.com.

**Makarand Paranjape** started his teaching career at the University of Illinois at Urbana-Champaign as a teaching assistant in 1980. After moving back to India in 1986, he taught at the University of Hyderabad, as fellow, lecturer, then reader. In 1994, he transferred to the Indian Institute of Technology in New Delhi as an associate professor of humanities and social sciences and was later invited to apply for the English professorship at Jawaharlal Nehru University in 1999, where he has been since then. He is a founding trustee of Samvad India Foundation, a nonprofit, public charitable trust, and editor of *Evam: Forum on Indian Representations*, an international multidisciplinary journal on India. In 2004, he was awarded an LSI through Samvad India Foundation for his proposal on "Science and Spirituality: The Delhi Dialogues."

**Paul Swanson** is a specialist in Buddhist studies. He joined the Nanzan Institute for Religion and Culture in 1986, starting as an associate editor and then becoming a permanent fellow in 1990. He has been editing the *Japanese Journal of Religious Studies* for more than twenty years, and has been actively involved in the editing and publication of numerous books on Japanese and Asian religions (including a six-volume series for Asian Humanities Press). He has been director of the Nanzan Institute since April 2001 and acting director since 1999.

**Dr. Ryusei Takeda** has spent most of his academic life researching the Pure Land Buddhist teachings of Shinran, Western philosophy, Buddhist thought, interreligious dialogue, and religious pluralism. He has published a doctoral dissertation on Shinran's Pure Land Buddhism and Nishida's philosophy and comparative studies on Whiteheadian and Mahayana Buddhist philosophy and the philosophies of Nagarjuna and Vasubandhu. Subsequently he was invited to give an address at the Silver Anniversary International Whitehead Conference in 1998, which was attended by three hundred scientists, physicians, economists, and scholars of religion, including three Nobel Prize winners. He is presently director of the Center for Humanities, Science and Religion at Ryukoku University in Kyoto.

**Dr. Jacek Tomczyk** has been working as a lecturer in anthropology at the Cardinal Stefan Wyszynski University since 2003. That same year, he became head of the Institute of Anthropology. He is member of the editorial board of Studia Ecologiae et Bioethicea and presented papers during IX International Philosophical Congress in Istanbul (Turkey) and the International Anthropological Congresses in Zagreb (Croatia) and Komotini (Greece).

# Index